SCIENTIFIC AMERICANS

SCIENTIFIC AMERICANS

INVENTION, TECHNOLOGY, AND NATIONAL IDENTITY

SUSAN BRANSON

CORNELL UNIVERSITY PRESS
Ithaca and London

First published 2021 by Cornell University Press

Printed in the United States of America

Library of Congress Cataloging-in-Publication Data

Names: Branson, Susan, author.
Title: Scientific Americans : invention, technology, and
    national identity / Susan Branson.
Description: Ithaca [New York] : Cornell University Press,
    2021. | Includes bibliographical references and index.
Identifiers: LCCN 2021032218 (print) | LCCN 2021032219
    (ebook) | ISBN 9781501760914 (hardcover) |
    ISBN 9781501760921 (pdf) | ISBN 9781501760938 (epub)
Subjects: LCSH: Science—Social aspects—United
    States—History—19th century. | Technology—Social
    aspects—United States—History—19th century. |
    Science—Social aspects—United States—History—
    18th century. | Technology—Social aspects—United
    States—History—18th century.
Classification: LCC Q175.52.U5 B73 2022 (print) |
    LCC Q175.52.U5 (ebook) | DDC 303.48/30973—dc23
LC record available at https://lccn.loc.gov/2021032218
LC ebook record available at https://lccn.loc.gov/2021032219

For Mark

CONTENTS

SCIENTIFIC AMERICANS

Introduction
The Role of Science and Technology in the Creation of American National Identity

The United States had a lot to prove in 1776. Independence meant doing without two things Americans had taken for granted as colonials: an economy and an identity. Both were derived from the most important global power of the day. Liberated from the constraints of the colonial system, but at the same time handicapped by the withdrawal of Britain's all-encompassing trade networks, the nation began to find its way toward a new economic system. Developments in science and technology helped the United States achieve economic independence. Exploring the ways Americans chose to characterize discoveries, inventions, and mammoth civic projects illuminates how men and women articulated their belief that the United States was a rising empire. In other words, science and technology helped Americans create a new identity.

Postrevolutionary America was not radically different from colonial America. Farmers farmed, merchants pursued business, and everyone worried about water supplies, fuel, and traveling on bad roads. The ideas, activities, and events chronicled in this book are situated within the stuff of daily life. Nation and empire were concepts that preoccupied some people, but others not at all. While ideology does not put food on the table, or money in the pocket, it can be a rallying point for a collective endeavor to achieve those quotidian goals. That is why national identity and nationalism are central to my inquiry. John Murrin characterized American national identity as "an extremely fragile creation of the Revolution."[1] Tracing British colonials' engagement with science and technology as they made the transition to citizens of the United States illuminates how this initially fragile identity took on a more solid shape and purpose over the course of the first seven decades of the republic. Science and technology move from the periphery to the center of American activities when viewed as the foundation for American development.

Though united by a political system, Americans voiced their differences more often than they expressed consensus. But on one point there was general agreement: the necessity to develop and expand productivity, markets, and trade. Promoting the interests of the United States often focused on the scientific and technological achievements that enabled the nation to compete with its rivals. The way forward lay through government initiatives, private organizations, and individual enterprise. As Benjamin Park suggests, nationalism was "an ideological instigator for policy and action."[2] Yet this book does not employ a whiggishly progressive framework. While recognizing that a rhetoric of progress was deliberate and purposeful, I also draw attention to the high price paid by Indigenous people and enslaved African Americans. The drive for agricultural improvement was fueled by treaties that ceded native land. Slavery went hand in hand with both settler expansion and new technologies; the textile industry in the northeast relied on slavery. Cotton was the product the United States was most eager to develop both for domestic markets and for export. Expansion into the Southeast was on land formerly belonging to Native Americans. White settlers there grew cotton produced with improved plows, processed with the newly developed cotton gin, and shipped down the Mississippi via steamboat. Almost all this work was performed by enslaved men, women, and children. National identity was, in part, shaped in opposition to what Americans were not. Science was employed to reinforce a racial hierarchy. It became a tool to confirm racial difference and justify slavery. Science was also used to validate the federal government's civilizing program in the 1790s, the removals of Native Americans from their traditional homelands, and the wars of extermination.

The rhetoric linking improvement with nationalism derived not from the political disunion in 1776 but from a consumer revolution that began in the early eighteenth century. This consumer boom had reached historic proportions by the late eighteenth century.[3] In British colonial America, imported goods increased almost eightfold between 1700 and 1773, rising 43 percent between 1768 and 1772. American products exported to Britain also rose at a staggering rate, increasing by 64 percent between 1756 and 1776.[4] When it came to taste, Americans took their cues from Britain. They enjoyed

Staffordshire ceramics, silver, and silk fabric. But by far, the greater number of imported items purchased by colonials were those needed for farms, businesses, and households. The inventory of a Connecticut store owner, Jonathan Trumbull, illustrates the variety of goods produced in Britain and consumed in America. His customers purchased gunpowder, paper, pots and pans, pails, needles, knives, thimbles, buckles, buttons, combs, spectacles, nails, sewing silk, wire, and pewter dishes.[5] These items were necessities, not luxuries.

But luxury goods, too, were in demand. Consumer desires outstripped availability. Rather than forgo imported products, Americans devised ways to make their own. This was not as challenging as might be supposed. As David Jaffee has shown for rural New England, artisans contrived home-grown luxury goods, such as clocks and furniture, to meet local demand. As the rural gentry acquired incomes that enabled them to cultivate a taste for finer things, urban craftsmen moved into the countryside to cater to this market. Benjamin Cheney, a Connecticut clockmaker, is one example of American ingenuity in the decade before independence. Cheney's down-market clients wanted luxury goods but not at luxury prices. To avoid using costly imported metals, Cheney created a workable substitute: he built clock mechanisms entirely out of wood. Cheney's clocks were wound by pulling on rope weights, rather than using a key, but some of his clock faces were painted with false winding holes to mimic the more expensive, imported, metal eight-day clocks.[6] Cheney was what Joyce Appleby describes as an American artisan-capitalist. Cheney, and others like him, found solutions to pre- and postrevolutionary scarcity. These inventors, producers, and consumers were the promoters of science and technology in the early republic.[7]

Few of these individuals thought of themselves as scientists or technicians. Only a handful of major innovations before the twentieth century depended on scientific theory. Instead, they were derived from technical knowledge and skills. Movable type, the mechanical clock, guns, and the steam engine were the inventions of craftsmen, not scientists. Most people still used the term *natural philosophy* to describe inquiries (theoretical rather than practical) into natural laws. Natural philosophy also referred to a body of knowledge based on systematic investigation involving precision and objectivity. Air, electricity, and miniature creatures (seen for the first time thanks to the microscope) were all subjects of inquiry by natural philosophers. But the term *science* was increasingly used by the beginning of the nineteenth century. Benjamin Silliman's *American Journal of Science*, begun in 1818, was among the first publications that did so. Similarly, the word *technology* was not often employed before Jacob Bigelow's *Elements of Technology* was published in 1829. Appointed to the Rumford professorship in the Application of Science to the Useful Arts at

Harvard College in 1815, Bigelow held the first faculty position dedicated to instructing students in "the utility of the physical and mathematical sciences for the improvement of the useful arts, and for the extension of the industry, prosperity, happiness, and wellbeing of society."[8] Technology, defined by Bigelow as science that was useful knowledge, had come into its own. Practical applications of mathematics, physics, earth sciences, and biological sciences all supported the rising nation.

This book benefits from historical inquiries on three interrelated topics: (1) the promotion of scientific education and practices among non-elites, (2) the place of science and technology in American culture, and (3) the development of nationalism and national identity in the early republic and antebellum eras. Historians have increased our awareness of how non-elites, especially in Britain, participated in the project of the scientific enlightenment in the eighteenth century. Scholars more recently have turned their attention to the interrelationship between the production and consumption of scientific knowledge. James Delbourgo's work on the popular reception of electrical experiments in colonial America, Susan Scott Parrish's investigation of the intense curiosity eighteenth-century Americans had about the natural world, and Londa L. Schiebinger's work on botanical explorations are just a few examples of the scholarship linking ordinary people to scientific pursuits.[9] This book furthers these claims about popular involvement with science by demonstrating how widespread participation altered an elite narrative that excluded the majority of Americans from engagement with science and, more importantly, ignored a growing necessity: popular support for funding state and federal projects. Science and technology were part of the transformation of public culture in the eighteenth and early nineteenth centuries. By investigating what Simon Schaffer describes as entrepreneurial culture, historians have uncovered a multiplicity of knowledge producers, audiences, and venues that show science becoming part of the commercial public sphere. More recent work by Fred Nadis, Ralph O'Connor, and Paul Semonin document these public activities in the early United States.[10] Building on these studies, I explore the variety of venues in which Americans developed new practices that expanded the base of scientific education and encouraged greater participation in the production of scientific knowledge and its practical applications. Consumer culture provides evidence of this public engagement with the products of discoveries and new technologies. I emphasize the tangible and visual ways that men and women consumed new technologies and scientific ideas, and how they celebrated small successes as well as grand achievements. Men and women valued scientific knowledge and education as necessary to individuals, communities, and the nation. The drive for national development expanded opportunities for Americans to participate in the creation and diffusion of scientific and technical expertise. Because scientists, engineers, and technicians depended on private investors, municipalities, and the public at large to fund their projects, they relied on, and therefore encouraged, widespread interest in scientific discoveries and technologies. Interdependency benefited all these constituencies.

Chapter 1, "Domestic Science," sets the stage for the activities related to science and technology in postrevolutionary America. It explains how informal scientific education provided by almanacs, public lectures, and demonstrations along with the financial encouragement of early scientific societies, generated an enthusiasm for the application of science and technology to civic, commercial, and domestic improvements, first in the colonial era, and then in the early years of the republic. Not all were welcome to contribute to this goal; yet women and all people of color, as outliers to citizenship, nevertheless made their presence felt. Scientific knowledge was within the grasp of many who were not white, affluent, and male.

Chapter 2, "Flights of Imagination," explores the link between air balloon technology and national ambitions—economic, political, and martial. Enthusiasm for the new technology brought large crowds to view launches, and inspired poetry, fiction, and consumer items. More than simply novelty, air balloons became a necessary component of fairs, parades, and other celebrations as an icon of progress. The most significant interactions with aerial technology occurred in the antebellum era when balloons became a form of commercialized leisure as well as an expression of national power. By the 1860s, they were weapons of war. And as mascots of imperialism, balloons demonstrated sovereignty over Native Americans.

Chapter 3, "Engines of Change," explains why machines captured the public's attention in so profound a way by surveying the early nineteenth-century venues in which Americans encountered mechanical technologies. Demonstrations of new machines acquainted men and women with inventions that enhanced daily life. One continually pursued idea was resurrected in an American context when Charles Redheffer's perpetual motion device caught the public's imagination. It sparked a national discussion of how the invention would propel the United States to global preeminence. Redheffer's failure to deliver on his promise dampened, but did not diminish, American hopes that technology was the answer to national development. The steam engine, on the other hand, quickly became useful and ubiquitous. The engines that propelled boats and railcars and powered looms and spindles literally drove the changes that helped the United States strive for economic independence. Americans expressed their enthusiasm for the steam engine's importance to national prosperity in poetry, fiction, and the purchase of steam-themed consumer items.

Chapter 4, "Grand Designs," illustrates how civic projects articulated national ambitions. The United States needed technologies to develop industry, transport American-made products, and protect the health and homes of urban citizens. But it was the way Americans chose to promote inventions, devices, and civic constructions that put national aspirations into public conversation. Although the Erie Canal was the most prominent public works project in the early nineteenth century, water systems undertaken in American cities between the 1810s and 1850s were the most tangible evidence of civic development. Celebrated as representations of America's place among the nations of the world, the magnitude of these projects made them tourist attractions. Lithographs, paintings, sculpture, and music depicted these achievements. Moreover, the brick-and-mortar constructions visually articulated empirical ambitions through their design and embellishment. Reservoirs, pumping stations, and fountains incorporated references to ancient empires and communicated Americans' conception of nationhood with an easily understood visual rhetoric.

Chapter 5, "Internal Improvements," explores how phrenology enabled Americans to apply its principles to key issues of the antebellum era. Riding the crest of reform movements, this science was a method for Americans to improve themselves. Phrenology's presence in popular culture acquainted men and women with theories of the human mind. It also gave Americans a new vocabulary with which to discuss important issues of the day; the science could be a weapon for satire, or a tool to validate or challenge long-standing assumptions, especially about race. Although Native Americans received attention and study from phrenologists, for the majority of Americans, Indigenous people were distant and unseen. If white Americans thought about native peoples at all, it was as a barrier to western settlement. Reports on ancient American artifacts often interpreted remains of Indigenous civilization as belonging to an extinct people—not the living owners of western land. People of color, on the other hand, were highly visible. Whether in town or country, white and Black Americans encountered each other daily. In the antebellum era, this familiarity with men and women of a different race encouraged both proslavery advocates and abolitionists to employ phrenology to serve their cause.

Chapter 6, "Fair America," shows how the exhibitions and contests organized by scientific societies in the first half of the nineteenth century fostered

American technological development. The Franklin Institute for the Promotion of the Mechanic Arts in Philadelphia and the American Institute of the City of New York for the Encouragement of Science and Invention were the most active societies in this era. The Exhibition of the Industry of All Nations at the Crystal Palace in New York City staged the largest industrial exhibitions in the early nineteenth century. All three organizations pursued a strategy of fairs and competitions to promote American products and to encourage invention and innovation. Fairs were a spur to economic competition, national prosperity, and American supremacy in global trade—goals that were shared by organizers, participants, and visitors.

The American embrace of discoveries and devices reveals enthusiasm for improvements small and large, personal and national. Exploring the many ways in which Americans expressed this enthusiasm reorders the historical focus on the first decades of the nation. Rather than viewing science and technology through the lens of political and economic development, this book reverses the focus in order to demonstrate the symbiotic relationship between political economy and innovation. One did not exist without the other. The urgency with which Americans pursued economic development reflected both necessity and ideology. Framing development as a national mission made new inventions and discoveries a necessary component of this endeavor. Economic initiatives sponsored by societies and civic groups provided opportunities to articulate national self-definition. New sciences and technologies were the tools used to characterize an emerging national identity.

Timothy Dwight, president of Yale College, expressed this national mission when, after traveling eighteen thousand miles over a ten-year period throughout the northeastern United States, he described the nation as a place "where regular government, mild manners, arts, learning, science, and Christianity have been interwoven in its progress from the beginning." This, he asserted, "is a state of things of which the eastern continent, and the records of past ages, furnish neither an example nor a resemblance."[11] For Dwight, independence from Britain did not disrupt the practices and initiatives Americans had long engaged in; a continued emphasis on technological know-how, fostered during the colonial era, powered the country's growing industrial and economic strength. Yet, at the same time, Dwight identified uniquely American characteristics. If, as Kariann Yokota suggests, Americans had to unbecome British in the first decades of the republic, one way of doing so was to construct an identity based on practical needs. Americans had to invent the nation.[12]

CHAPTER 1

Domestic Science

*Learning, Observing, and Promoting Science
as American Enterprise*

Americans have always been science-minded.
As colonists, men and women observed, experimented, and tinkered. People of all ranks in British America were involved with scientific ideas and practices. From necessity or simple curiosity, colonists were busily learning, observing, and entertaining themselves through engagement with the natural world.[1] The aspirations expressed in the postrevolutionary era about the nation's development were grounded in centuries-old practices and activities. After independence, a ready-made, educated audience was motivated to encourage progress through innovation. Americans possessed the means, motive, and opportunity to gratify their curiosity and to acquire knowledge. The means to do so were ready to hand: formal and informal modes of learning, social gatherings, and enthusiastically attended events. Motivation combined the personal with the political: the legacy of revolution gave Americans a language with which to articulate ambitions for national self-sufficiency, the expansion of trade, and internal development. Within this national framework, individuals sought improvements for themselves, their families, and their communities. As Joyce Appleby asserts, the first generation of Americans reared in the postrevolutionary era had a new sense of nationhood, one in which public prosperity depended on personal initiatives. Personal progress was national progress.[2]

This chapter explores opportunities for learning about science and technology by situating those activities within a context of nationalism—first as subjects of Britain and then as citizens of the newly independent United States. The nation's future depended on its citizens to improve agriculture, manufacturing, transportation, and even people. The foundation for this postrevolutionary agenda was laid down in the colonial era; the goals and ambitions for colonial contributions to the success of the British Empire became the goals and ambitions to develop an American Empire.

Membership in the British Empire defined the commercial lives as well as the political lives of American colonials. Trade with Britain dictated the terms of economic activity, including the products that colonials could manufacture. Colonial life combined dependence with independence: the necessity to purchase manufactured goods from Britain, and the production of homegrown goods both for domestic consumption and for the empire. The term *improvement* was used by everyone for almost everything connected with domestic economy. Efforts to develop land, extract resources, and make products for domestic use and for trade pushed technological development. An expanding print culture enabled the transmission of information through books, periodicals, almanacs, and newspapers.

Improvement was a collective endeavor and education was one means of achieving this goal. William Penn demonstrated his ambition for the success of his colony by offering premiums "to authors of useful sciences and laudable inventions."[3] Half a century later, Benjamin Franklin continued Penn's vision with a plan for public education designed to foster the skills and talents needed for scientific and technological development among Pennsylvanians. Franklin's *Proposals Relating to the Education of the Youth in Pennsylvania* (Philadelphia, 1749) stressed practical knowledge in drawing, arithmetic, accounts, geometry, astronomy, and agriculture. When the Academy of Philadelphia opened in 1750, surveying and navigation were added to Franklin's original list.[4] Though Franklin had briefly attended Boston's Latin School, which put education within the reach of Boston's artisan class, his ambition for Philadelphians in the second half of the eighteenth century was to acquire knowledge of a practical kind. This was something he had achieved only after leaving formal education behind. Franklin believed that the primary purpose of education was self-improvement, but it also enabled individuals to serve their country.[5] Franklin was not alone in promoting practical education to achieve national ends. King's College in New York (founded in 1754) taught surveying, geography, husbandry, commerce, government, and natural sciences. Samuel Johnson, then president of King's College, believed that education should prepare young men to be "useful to their country in public stations when they come forth to act their parts upon the stage of life."[6]

Yet most colonials learned by doing. Ambition to be recognized as participants in the empire drove their engagement with the natural world. Cadwallader Colden in New York, John Bartram in Pennsylvania, and Alexander Garden in South Carolina, to name a few, eagerly engaged with the empirical networks of knowledge centered on Britain and the Royal Society. They shared observations and specimens with correspondents in London who were eager for information about the colonial environment. In Charleston, South Carolina, the nursery owner Martha Logan procured plants for customers in the colonies and in London. When John Bartram met Logan during his southern tour in the 1760s, he paid her the compliment (in a letter to Peter Collinson rather than to Logan herself) that she knew her horticulture well. Logan's own letters confirm how knowledgeable she was, and they demonstrate that she was part of a local network of fellow horticulturists. Logan's *Gardener's Calendar*, which instructed southern gardeners about what, and when, to plant, was printed at the back of South Carolina almanacs until the end of the eighteenth century.[7] Another colonial woman, Jane Colden, diligently studied the Linnaean system (albeit in English rather than Latin). As a teenager living with her father, the scientist Cadwallader Colden, in New York's Hudson Valley in the 1750s, she put her knowledge into practice by classifying over three hundred species of local plants. The London collector Peter Collinson learned of Colden's achievements through correspondence with her father, and Collinson passed on to Linnaeus himself the fact that Jane Colden was "perhaps the first lady that has perfectly studied Linnaeus' system."[8] In Pennsylvania, Susanna Wright raised silkworms. She shipped raw silk to London where it was spun into cloth. Wright shared her knowledge and practices in her essay, "Directions for the Management of Silk Worms."[9] In South Carolina, Eliza Lucas Pinckney experimented with a strain of West Indian indigo that would grow in South Carolina. The success of her efforts guaranteed the colony, and British textile manufacturers, a profitable commodity for decades to come.[10] She was also a sericulturist. Pinckney's slaves produced silk that was presented to the Dowager Princess of Wales as a tribute from colonials, but also to demonstrate a colonial success story: Americans produced valuable goods that the mother country could not.[11]

On the "Page of Improvements" in Daniel Leeds's 1712 edition of his *American Almanack*, he urged young women and young men to plant gardens and orchards. As an incentive to local efforts, he offered his recipe for hard cider. Leeds urged his fellow colonists to rely more on domestic produce and less on imported "Forreign Liquors, Rum, Wine, and the like." Leeds saw this activity not just in terms of cost effectiveness; gardens and orchards (and cider) contributed to "the publick good and improvement of our Country, as well as particular Interest."[12] This thread of nationalism was woven into education, instruction, and domestic economy. Americans received this message in a variety of venues, including, most importantly, print culture.

The Diffusion of Information

Most British colonials engaged in networks of knowledge conveyed through newspapers, magazines, and almanacs. Benjamin Franklin's electrical kite experiment, for example, was first conveyed to the public in the newspapers.[13] Thanks to a postal system that linked the British colonies, newspapers reached communities from Massachusetts to Georgia. There were twelve English-language newspapers in 1735 and seventeen in 1760. By the eve of the Revolution, the number had grown to forty, making scientific information available to almost every literate American.[14] In 1728, for example, the Philadelphia printer Samuel Keimer announced that his forthcoming paper the *Universal Instructor* would include essays and information about the sciences, subjects that Keimer considered to be "the richest Mine of useful Knowledge."[15] As print culture and Atlantic trade networks expanded in the second half of the eighteenth century, colonials had increasing access to books, pamphlets, and objects sent from Britain. Booksellers in Boston, New York, Philadelphia, and Charleston advertised how-to manuals for tradesmen and farmers, scientific texts, and catalogs of instruments. Subscription libraries such as the Library Company of Philadelphia and the Charleston Library Society purchased instruments along with the latest scientific texts.[16] Coffeehouses offered customers local and imported newspapers with their caffeine and tobacco. Small circulating libraries often shared space with other businesses frequented by both men and women. Lewis Nicola's Philadelphia library shared space with a milliner. This accommodation made books and periodicals readily available to his female customers.[17] By the end of the colonial era, Americans were producing their own periodicals. The *Pennsylvania Magazine, or, American Monthly Museum* (January 1775–July 1776) was the last to appear before independence. Edited by Thomas Paine, the magazine was intended to provide a general audience with "instruction and information."[18] Information included news of recent American inventions such as a fire escape, dredger, sulfur furnace, and waterproof cement. Articles were sometimes accompanied by engraved plates of the devices, making it easier for readers to understand, and possibly to replicate, the technologies described. Recognized scientists such as the astronomer David Rittenhouse and the physician Benjamin Rush were among the magazine's contributors.[19] In the context of political upheaval and the real possibility of disunion, Paine's stated goal for the publication was "to contribute every thing in our power towards the improvements of America." Public promotion of American invention was a first step in linking science to an emerging national identity.[20]

Men and women without access to libraries, or the money to purchase books and magazines, read snippets of the most prominent natural philosophy writings in almanacs. The poor man's science textbooks, almanacs were more widely available than either magazines or newspapers; they were in almost all European American homes by the mid-eighteenth century. Nathaniel Ames's *Astronomical Diary and Almanack* reached an estimated sixty thousand New England readers in the 1750s. A storekeeper in Leominster, Massachusetts, for example, sold enough almanacs to supply every household in his area. And that was just one shop in one town. Even in the far less densely populated southern colonies, the printers Hunter and Royle of Williamsburg, Virginia, sold between four and six thousand almanacs a year.[21] Moses and Graham Parsons are examples of how this almanac information was valued: the brothers cut out items of scientific interest and interleaved them with the pages of their diaries.[22]

Almanacs were not purchased and then tucked in a drawer, to be occasionally retrieved and consulted. They were kept in plain sight and readily at hand in kitchens and offices. Many surviving copies have a hole in an upper corner where they hung from a nail. A few still possess a loop of string, an indication of their location within easy reach for daily needs. Almanacs were a constant companion of farmers, merchants, fishermen, lawyers, travelers, and just about everyone in British America. Cheap, useful, and portable, almanacs conveyed all sorts of information to readers. Among the typical contents of a colonial almanac were astronomical calculations; tide tables; a list of court days for the region (a Boston almanac such as Nathaniel Ames's included the courts in Massachusetts, New Hampshire, Rhode Island, Connecticut, and Vermont); mileage calculations for roads going north and south from the principal city; fairs for the region; a table calculating simple interest; and a currency converter—from French pistoles to British guineas to Spanish dollars. Nathan Bowen, author of *The New-England Diary, or Almanack*, was not exaggerating when he told his readers in 1733 that almanacs "are of absolute necessity in the Business and Affairs of Life."[23]

Almanacs remained an important resource in postrevolutionary America. Farmers, businessmen, travelers, and pretty much anyone who needed to know the tides, lunar cycles, or mileage between towns, or wished to learn snippets of history or science, relied on almanacs for information. In order to rise above the competition as the number of almanac makers increased in the first decades of the republic, they advertised that superior calculations set them apart from their fellows. But what really distinguished one from another, or what made one almanac more appealing than the rest, was the additional information, entertainments, or items of special interest it contained. *Benjamin Banneker's Pennsylvania, Delaware, Maryland and Virginia Almanac* had all these elements. Banneker was a surveyor, mathematician, and almanac maker. He was also African American.

Banneker was largely self-taught. In the late 1780s, he borrowed books from his neighbor George Ellicott, an astronomy enthusiast. Banneker quickly mastered the complicated mathematics necessary for calculating celestial movements and produced an ephemeris for 1791. Many of the instruments needed to calculate celestial movements were also employed to measure land. Thus Banneker's success with this first publication led to employment on the survey of the city of Washington.[24] His almanacs, published from 1792 to 1796, were a commercial success, not only because they were painstakingly accurate but also because Banneker used his publications as a platform to argue against slavery.[25] Banneker did not hide his race. The almanac's editors proclaimed the work to be "an extraordinary Effort of Genius . . . calculated by a sable Descendant of Africa, who, by this Specimen of Ingenuity, evinces, to Demonstration, that mental Powers and Endowments are not the exclusive Excellence of white 'People, but the rays of Science may alike illumine the Minds of Men of every Clime, (however they may differ in the Colour of their Skin)."[26] Lest readers remain unconvinced of Banneker's talents, the editors assured them that America's foremost man of science, David Rittenhouse, personally confirmed the accuracy of Banneker's calculations.[27]

Banneker's most important target was Secretary of State Thomas Jefferson. Jefferson's views on the inferiority of Blacks, expressed in *Notes on the State of Virginia*, were well known. Banneker sent his ephemeris for 1792 to Jefferson. The secretary's reply (which was later published by Quaker abolitionists) admitted Banneker's skills: "Nature has given to our black brethren, talents equal to those of the other colours of men." Jefferson's comment appeared to be a revision of his ideas on racial difference. What seemed to be mental inferiority in African Americans, he wrote, was "owing merely to the degraded condition of their existence both in Africa and America." Moreover, Jefferson did what he could to broadcast Banneker's demonstration of the mental equality of whites and Blacks by sending Banneker's almanac to the Marquis de Condorcet, secretary of the Academy of Sciences.[28]

But privately, Jefferson was less generous. In a letter to Joel Barlow, Jefferson suggested that George Ellicott may have aided Banneker in his calculations.[29] Despite Jefferson's refusal to acknowledge Banneker's talent, the almanac maker had outstanding intellectual abilities that set him far above most people. His background and upbringing, among a small community of free people in Maryland, illustrate one of the rare opportunities for men and women of color to publicly participate in American science.

Celestial Connections

Astronomical calculations were the most prominent information in almanacs; sun and moon risings and settings, moon phases (and the moon's connection to tides), and the appearance of comets and eclipses aided farmers, travelers, sailors, and scientists. Before the Revolution, this information also provided a tangible link between colonials and Britain; Bostonians, Charlestonians, New Yorkers, and Londoners viewed the same skies and collectively engaged in sharing observations across the British Empire. Especially in the year of a rare event, such as the second transit of Venus in 1769, almanacs delivered information and instruction. Readers obtained, in digest form, the same astronomical education to be had in expensive textbooks. The *New Jersey Almanack* for 1769, authored by "Copernicus Weather-Guesser," for instance, contained a dialogue between an astronomer and a young lady. This was a common format in many science texts; instruction was framed as dialogues, conversations, or catechisms between fictional characters. This method made it easier, and perhaps more entertaining, for readers to understand the concepts presented to them. The 1769 *New-England Almanack, or Lady's and Gentleman's Diary* contained information about the upcoming transit of Venus, including a history of transits going back to the sixteenth century.[30] Even the orrery, a model of the solar system, was within the reach of almanac readers, who were assumed to have a level of interest and understanding of astronomy. *Poor Richard Improved* provided basic information about the solar system for those "who have not Leisure or Opportunities for reading Books of Astronomy."[31] Benjamin Franklin's almanac for 1753 described a solar eclipse and explained the earth's axis and elliptic plane in relation to the sun. Franklin suggested that "any Person who has not an Opportunity of seeing an Orrery, may easily represent this by an Apple or other round Body with a Wire thrust through the Middle of it."[32] In a letter to the *Boston Gazette* in 1757, one reader complained that two local almanacs completely contradicted each other about a coming eclipse—evidence that readers paid attention to an almanac's accuracy.[33]

Whether they were embracing new ideas about the natural world, or adhering to old beliefs, celestial events mattered to eighteenth-century British Americans. One tangible piece of evidence that men and women were keen observers of the skies is a medal unearthed during the excavation for the National Constitution Center in Philadelphia. The area was once a bustling, crowded neighborhood, and the token may have dropped from someone's pocket as they did their marketing or walked down Chestnut Street to a tavern. The medal is a calendar for the year 1758. Often given away by merchants as reminders of their wares, medals showed the dates for all the Sundays in the year and the beginning of each legal term for the courts— Michaelmas, Hilary, Easter, and Trinity. This particular medal, because it was for 1758, bears the inscription "This year expect the comet without danger."[34]

The astronomer Edmund Halley was the first to correctly identify the orbit of the comet that passed through the earth's atmosphere roughly every seventy-five years. Halley's first prediction for the comet's return was 1758.

FIGURE 1.1. Calendar medal for 1758. Courtesy of Independence National Historical Park.

Its expected appearance was well publicized and important enough to be noted on the coin.

Philadelphians did not need a reminder that the comet was due, but they may have sought reassurance, such as that expressed on the coin, that it was not going to vaporize them. Long-held beliefs that celestial events were signs, omens, or punishments were gradually giving way to a rationalist perspective on nature. But lingering doubts, as the calendar token shows, remained. The science author and demonstrator Benjamin Martin complained about the ignorance of some of his audiences: "There are many places I have been so barbarously ignorant that they have taken me for a magician, yea, some have threaten'd my life for raising storms and hurricanes."[35] The transition to the Newtonian view of the universe was gradual. The *Virginia and Maryland Almanac* for 1732 described eclipses in a straightforward, technical manner but still connected them with natural disasters.[36] Nor did the new science abandon God. Appreciation and exploration of the Book of Nature formed the basis and rationale for scientific inquiry. Belief in astronomical phenomena as divine messages (and often judgments) was still evident in 1758: Richard Draper's *Blazing-stars messengers of God's wrath* cautioned both saints and sinners about how to behave themselves "when God is in this wise speaking to them from heaven."[37] Whatever interpretation individuals put on celestial phenomena, colonials possessed the necessary information to prepare for eclipses, comets, and transits across the sun.

Long before achieving independence in 1783, these practices prepared women and men to be citizens of a science-minded nation. But how people engaged with celestial phenomena largely depended on education and wealth. The most affluent colonials paid for lectures or demonstrations of the latest imported instruments. Colleges, attended by wealthy young men, owned all kinds of apparatuses for teaching natural philosophy. Harvard College, for instance, had arguably the finest collection of astronomical instruments on the continent, largely due to the efforts of John Winthrop, the Hollis Professor of Mathematics and Natural Philosophy. Institutions such as the American Philosophical Society, the Library Company of Philadelphia, and the Charleston Library Society also owned high-quality instruments commissioned from British makers. Only a handful of colonials possessed the skills necessary to construct precision instruments. David Rittenhouse was one such craftsman. In addition to clocks and telescopes, Rittenhouse constructed two large orreries —one for Princeton University and the other for the University of Pennsylvania.[38]

Armed with astronomical knowledge and forewarned of impending events, Americans observed the heavens. Comets could be seen with the naked eye; total solar eclipses noticeably darkened the sky. Many wealthy colonials purchased the proper equipment to observe transits, comets, and eclipses.[39] Probate inventories list telescopes and spyglasses (spyglasses were more often used for sighting points on land or at sea, but they did magnify objects in the night sky). George Johnston's probate—recorded in Fairfax County, Virginia, in 1767—listed two telescopes. He also had an "Electrifying machine," indicating that Johnston was deeply invested (at least financially) in both astronomy and experimental science.[40] Almanacs hastened to assure less affluent readers that cheap alternatives to spyglasses and telescopes were available—the sun could be safely viewed through a pane of smoked glass.[41]

Men, women, and children all over North America and the West Indies craned their necks to look at the skies. Reports on the more notable celestial phenomena document the popularity of these events. After a 1722 solar eclipse visible in Boston, one newspaper acknowledged, "There were too many Spectators then, to make it [the eclipse] now a piece of publick News: The Hills and Turrets were crowded with gaping Planet-Peepers."[42] The transit of Venus in 1769 was the most publicized astronomical event of the eighteenth century. It was the second of a pair of transits (the first in 1761 was obscured by clouds on most of the North American coast). The transit was no ordinary solar eclipse: the crossing of Venus over the sun enabled astronomers to accurately measure the earth's distance from the sun and the relative distances of the other known planets. For the less astronomically inclined, accurate measurements of the transit meant a more precisely calculated longitude. And that data had economic, and possibly geopolitical, implications for everyone.[43]

As a consequence of the buildup to this not-to-be-missed event, almanacs and newspapers lavished page space on general astronomy, eclipses, and when and how to view the transit. Observations in North America were planned down to the last detail: multiple locations, a cadre of telescopes and other instruments, and special viewing stations. David Rittenhouse erected a building especially for the transit at his home outside Philadelphia. This celestial event, more than any other, left a record of participation—not just of the official observers, but of those who observed the observers. Hundreds of people crowded around the platform in the State House yard in Philadelphia. At Rittenhouse's property twenty miles away, there were so many people that the astronomers feared they would not be able to communicate necessary times and measurements because of the noise. Afterward, William Smith recounted his surprise and delight that the crowd remained so quiet that "there could not have been a more solemn pause of silence and expectation, if each individual had been waiting for the sentence that was to give him life or death."[44]

This activity was instructive for some and entertaining for many. It also reminded colonials of their ties to Britain. England was the home base for orienting the temporal and geographical British colonial world. All astronomical calculations referenced Greenwich longitude, latitude, and time of occurrence. Like the earth in a pre-Copernican universe, everything revolved around Greenwich. Almanac readers were often reminded of their place in the British Empire through a "retrospective simultaneity."[45] For example, Roger Sherman's New York almanac gave the exact hour that a solar eclipse would occur in May 1752.[46] New Yorkers knew when it would be visible to them, but Sherman also noted who else would see it: while a colonist in Manhattan was staring through her smoked glass, men and women in Jamaica, London, and Newfoundland were doing the same. Even when eclipses were invisible at a reader's location, almanacs assured readers that their fellow Britons had a view. Rhode Island almanac readers were alerted to the upcoming Transit of Mercury in 1753, an event only visible with the aid of a good telescope and a clear sky, but with luck, the almanac told readers, they would see "the Planet Mercury . . . rise in the Sun. . . . According to Dr. Hally [sic], the Middle of this great Conjunction will be observed in London at 20 Min. after 7 o'Clock."[47] Franklin told his almanac readers, "If you get up betimes, and put on your Spectacles, you will see Mercury rise in the Sun, and will appear like a small black Patch on a Lady's Face."[48] After the fact, the *Pennsylvania Gazette* noted that the transit, though obscured by clouds in the Philadelphia area, had been clearly observed by Britons in Antigua.[49] *Poor Richard Improved* that same year listed a lunar eclipse on April 17 that was not visible in North America, but noted "In London the Moon will rise five Digits eclipsed."[50] In June 1764, a Boston newspaper informed readers that colonials had shared their view of the recent solar eclipse with two of the royal princes, who graced the Greenwich observatory with their presence during the event.[51] Observation was geographically boundless: individuals physically separated by oceans, forests, and large chunks of the North American continent saw the same stars and observed the same celestial activities. Observation bound all Britons to one another—Massachusetts farmers, Glasgow merchants, Antiguan planters, and royal princes.

Viewers of transits in all the British territories, including the "curious in North America," were invited to share their data with other colonists and with the Royal Society.[52] Such data sharing was a way to become an acknowledged part of the British scientific community. Measurements of the transit of Venus taken by the American Philosophical Society members at

Philadelphia and Norriton were conveyed to Nevil Maskelyne, the astronomer royal. In turn, Maskelyne formally thanked the APS and encouraged future astronomical endeavors. The satisfaction these colonials took in such attention from an important scientist is reflected in the inclusion of Maskelyne's letter in the first volume of the society's *Philosophical Transactions*. And Maskelyne's thanks were no mere courtesy—the transit of Venus in 1769 was a global enterprise. National prestige was at stake (it was unthinkable that the French might take more accurate measurements than the British). The Royal Society requested members of the British Empire, wherever they happened to be, to submit observations of the transit. The society even provided Captain James Cook with an astronomer and a portable observatory to take measurements in the South Pacific.[53]

A few colonial observers derived a certain measure of fame from their contributions to British progress. The London *Gentlemen's Magazine* (a periodical available to Americans), mentioned reports on Halley's comet sent to England from North America and the British West Indies in 1758.[54] The pinnacle of ambition was achieved by a handful of colonial men whose observations were printed in the Royal Society's *Philosophical Transactions* (not to be confused with the *Transactions of the American Philosophical Society*). For colonial scientists, it did not get any better. Colonials were part of the project of British science because of their location, not in spite of it; colonials could do what Londoners could not and what colonials did was an important contribution to British scientific endeavors and to British prestige.

Organized Americans

One of the most enduring elements of American efforts to promote science and technology was the organizations designed to facilitate the spread of scientific knowledge and to encourage invention and improvement. A few, begun in the colonial era, weathered the Revolution and thrived in the early republic. These organizations aimed to promote the economy by financially encouraging farmers and mechanics. In New York, the newly formed Society for the Promotion of Arts, Agriculture, and Oeconomy, stirred by "a zeal for public Emolument," entreated "all lovers of their Country" to communicate experiments in agriculture and knowledge of mines and minerals.[55] The society advertised prizes for the best quality hemp, flax, apple seedlings, cheese, hops, and barley and for the production of potash, linen cloth, linen yarn, stockings, hides, shoes, deerskins, tiles, and slate.[56] Philadelphians formed two separate learned societies in the 1760s: the American Society held at Philadelphia for Promoting Useful Knowledge and the American

Philosophical Society.[57] Both groups set themselves multiple tasks: to be part of a transatlantic network of knowledge through correspondence with British and European learned societies (most especially the Royal Society of London) and to pursue practical solutions to everyday problems. The table of contents of the first volume of the *Transactions of the American Philosophical Society* testifies to the earnest search for improvements in agriculture, navigation, and commerce that would "tend to the improvement of their country, and advancement of its interest and prosperity." The society urged Americans to seek out natural products "for the support and ornament of life, and for articles of trade and commerce." Just as early nineteenth-century organizations, such as the American Institute, sought to replace foreign goods with domestic ones, colonials hoped to make a greater contribution to Britain's balance of trade. The American Philosophical Society emphasized that greater knowledge about, and use of, native plants and minerals would "render us more useful to our mother country."[58] Cultivation of American vines, for example, might produce wines for export as good as or better than wines from France. The society hoped that "in time, our wine might be much esteemed."[59] In particular, the society's mission statement took pains to compare the climatological and geographic similarities between Philadelphia and Peking, a circumstance that meant colonial America might supply the British Empire with domestic products such as porcelain, silk, ginseng, and tea. With this wish list in mind, the APS, at the encouragement of Moses Bartram, established the Society for Promoting the Cultivation of Silk in 1770.[60] These northern silk enthusiasts were late to join the silk endeavor in the British North American colonies. South Carolina already had a standing Committee on Silk Manufacture. In 1769 that colony exported 330 pounds of raw silk, worth £2,173, to Britain.[61] American silk was transformed into fabric by Spitalsfield weavers in London. Always the champion of colonial manufactures, Benjamin Franklin gifted American woven silk to parliamentarians and royals.[62] And some American silk found its way back across the Atlantic to clothe wealthy colonials.

To encourage new inventions, the American Philosophical Society extended a flattering request that "expert and ingenious" American mechanics send along "any new inventions and discoveries they shall make." A two-hundred guinea prize for the best submission sweetened the offer.[63] A lack of formal education was not a bar to successful invention and innovation. Although the majority of Americans had no instruction beyond reading, writing, and basic arithmetic, most men and women relied on skills gained through hands-on experience. Knowledge of chemistry, engineering, and materials science was acquired through problem solving rather than school-ing.[64] The variety of models for tools, instruments, and machines submit-ted to the APS in the late eighteenth century attests to Americans' abilities. Before the United States Patent Office began in 1790, individuals who sought to record that they were the bona fide inventor of a new device or process did so by sending it to this well-known and well-respected organization. In 1785 John Fitch, one of the first Americans to construct a steamboat, submitted a model of a steam-driven paddleboat. Other inventions, such as a model bridge, model stoves, fireplaces with improved dampers, a ship's pump, a machine for cutting files, and a horizontal windmill, found their way into the society's collection.[65]

Gendered Networks of Knowledge

Women, often outsiders to formal networks of knowledge such as the Amer-ican Philosophical Society, found other venues in which to pursue opportu-nities to become part of a scientifically informed community. Women were an acknowledged audience for natural philosophy. Educated women read and discussed the latest scientific works. They attended lectures and demon-strations, and visited the first American museums. Some French and British texts, such as Jean-Antoine Nollet's *Leçons de physique expérimentale* were spe-cifically addressed to boys *and* girls. Bernard Fontenelle's best-selling *Conver-sations on the Plurality of Worlds* was directed at young women. The lessons in Fontenelle's text are embedded within a fictional framework in which a young marquise discusses astronomy and other topics in natural philoso-phy with her male preceptor. Like Fontenelle's work, Pluche's *Spectacle de la nature* employed the conceit of fictional instructors (male and female). British readers with no knowledge of the French language could read the playwright Aphra Behn's translation of Pluche, or several other texts such as Benjamin Martin's *Young Gentleman and Lady's Philosophy*; James Fergu-son's *Easy Introduction to Astronomy for Gentlemen and Ladies*; Tom Telescope's *Newtonian System of Philosophy, Adapted to the Capacities of Young Gentlemen and Ladies* (written by John Newberry); and Elizabeth Carter's translation of Francesco Algorotti's *Sir Isaac Newton's Philosophy Explain'd. For the Use of the Ladies*. All these publications were available to colonials, either through personal purchase or early societies and libraries.[66]

As learning became part of a burgeoning leisure industry in the eigh-teenth century, women gained further access to science.[67] Public demonstra-tions of experiments led men and women to make purchases for private instruction and entertainment.[68] Science could even aid courtship; George Ellicott gave his future wife, Elizabeth Brooke, a copy of Ferguson's book.[69] The marketing of ideas and the sale of instruments went hand in hand. In the 1760s Joseph Priestley urged "all persons who propose to understand the subject of Electricity, though they have no expectations of making discoveries, to provide themselves with an electrical machine, or at least desire some of their friends to show them the experiments." These objects identified their owners as part of social as well as scientific networks.[70]

The abundance of information about, and access to, the latest scientific ideas had significant consequences for colonials. Bostonians, Philadelphians, Charlestonians, and residents of surrounding hinterlands purchased or borrowed imported texts and instruments, and listened to itinerant lecturers. Colonial men relied on informal epistolary networks of learning to obtain books and instruments from Britain, and to transmit discoveries and observations (such as astronomical data) to their Britain-based correspondents. Though excluded from formal institutions for scientific inquiry, women formed their own intellectual communities. In the Philadelphia area, Milcah Martha Moore, Hannah Griffitts, Susanna Wright, Elizabeth Norris, and Elizabeth Graeme Fergusson formed one such network. These women were scholars and authors—primarily of poetry and literary prose rather than natural philosophy—but their interests were wide-ranging.[71] Hannah Griffitts, for example, recorded her admiration for Pennsylvania's other homegrown man of science, David Rittenhouse, in verse form in 1776. But she cautioned the astronomer to stick to the stars rather than meddle in politics:

> Politicks will spoil the Man
> Form'd for a more exalted Plan,
>   Great nature bade thee rise,
> "To pour fair Science on our Age
> "To shine amidst th'historic Page,
>   And half unfold the Skies."[72]

Women recognized that scholarly achievements might seem beyond their sphere. Margaret Morris suggested in a tongue-in-cheek sketch how philosophy and science were aids to, rather than distractions from, household duties. When a pudding burst its bag, her cook

> explores the Pot, & behold, (O cruel fate!) the pudding Bag is burst!— down drops the Ladle—up goes her hands—she thought some dire Misfortune would befall 'em today, for two great Crows flapp'd at the Window—there's a Hobgoblin in the Pot.[73]

This unenlightened woman's response to the unfortunate pudding bag was derived not from science, but from superstition. Morris, on the other hand, calmly explained to her cook the chemistry behind the unfortunate accident:

> You have, said I (sweetly smiling) accumulated the Pablum too hasty upon the Fire, & by that Means, raised such a brisk Vibration & coalition among the ignited Particles thereof, which being communicated by the aquamedia (for this it is but susceptible of, & can only convey a certain Degree of heat, yet it will make a terrible Jumble in the Pot) to the component heterogeneous Particles of the Pudding, so as to extend their Bulk, & at the same Time rarify & disengage the latent Air, whose elasticity overcoming the tenacity of the Bag, & the tying thereof being too tight to give Way, a rupture in the weakest part of the Cloth, constituting the said bag, must happen of Course, & the Contents, *qua data porta ruint*, will rush out where they can get vent.[74]

Her amusing imaginary lecture was meant as a contrast to a woman who "has been taught to think it the highest absurdity to venture out of the domestic Province."[75]

Educated women such as Morris benefited from the transatlantic transmission of books and instruments. Circulating libraries openly welcomed female readers. Samuel Loudon's advertisement for his New York City library declared that "the ladies are his best customers."[76] Private libraries, such as the New York Society Library, Philadelphia's Library Company of Philadelphia, and the Union Library were open only to members, most of whom were men. But men borrowed books for female family members, as a Library Company member William Drinker did for his mother, Elizabeth. Whether in personal libraries or members-only libraries, women had access to the works of Pluche, Fontenelle, Newton, and others. A few of these works were generously illustrated to demonstrate and clarify principles and experiments. Volume 5 of Nollet's *Leçons de physique expérimentale*, for example, shows a woman gesticulating at the small wonders under a microscope.[77] Images as well as words reinforced the notion that women were legitimate participants in the pursuit of scientific knowledge.

New schools for young women offered a further opportunity to acquire scientific knowledge in the early American republic. In 1787 Benjamin Rush was the first to teach chemistry at the Young Ladies Academy of Philadelphia. His *Syllabus of lectures containing the application of the principles of natural philosophy, and chemistry, to domestic and culinary purposes* is remarkably similar in content to his lectures at the College of Pennsylvania. Rush's introductory remarks explained why chemistry was an appropriate subject for young

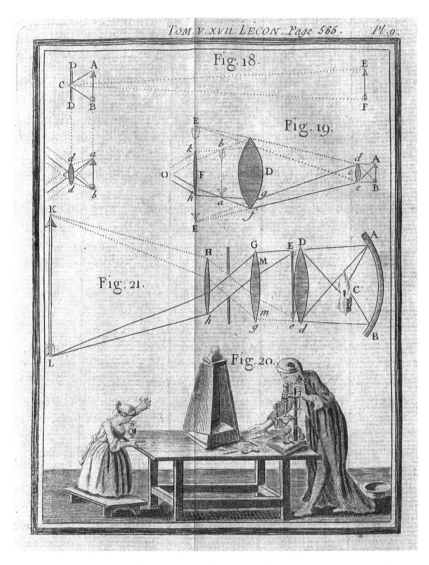

FIGURE 1.2. Jean-Antoine Nollet, *Leçons de physique expérimentale*, vol. 5, plate 9 (Paris: H. H. Guerin & L. F. Delatour, 1777). Library Company of Philadelphia.

women to study. If Rush is to be believed, chemical knowledge could perform wonders. It would, he explained to the young women, "excite a taste for such books as treat more fully upon these subjects, & raise you above the necessity of stooping to novels & romances for entertainment." It would also furnish them with "subjects for rational and improving conversation, and thereby preserve conversations from dress, fashions or scandal." It would, he told them, "cause your society to be sought for & courted by sensible men, & be the means of banishing fools & coxcombs from your company, afford you pleasure in solitude, and render you independent of public amusements for your happiness." And finally, chemical knowledge qualified them "to shine as wives & mothers—& mistresses of families when it shall please god to call you to fill those important female stations."[78]

The first section of his lectures was on general chemistry topics—heat, salts, earths, inflammable bodies, common metals, water, and air (including "dephlogisticated and phlogisticated airs"). Rush may have performed some chemical demonstrations to illustrate his points. This portion of the course seems to have been no different from that given to the students at the college. The second section of the *Syllabus*, however, was explicitly designed for young women. It dealt with the application of chemistry to the home—including building materials, fireplaces, insect and vermin control, heating and cooling, lightning protection, thermometers and barometers, food preservation, and care of the teeth. Such subjects, Rush assured his listeners, would make them "philosophical, as well as practical, housekeepers." Benjamin Rush initiated a tradition of chemical education at the Young Ladies Academy. Science was soon a common subject for young women everywhere.[79]

Twenty years after Rush introduced chemistry into the Young Ladies Academy curriculum, faculty member Benjamin Tucker offered lectures to the general public. In the broadside for his lectures, Tucker argued that chemistry lay within the female sphere, unlike other "walks of science," which "must be trod by men alone: the labour and attention required in their pursuit, as well as the objects to which they lead, exclude them from the literary province of females." But chemistry was appropriate because it "increases our knowledge of nature, but [also] gives us a noble display of the wisdom and goodness of its Author." Like earlier lecturers on electricity, Tucker included "appropriate and brilliant experiments" with his talks.[80] Tucker's course apparently had great appeal. He repeated his lecture series at least once, and then published *Grammar of Chemistry*, based on the lectures. His book included the subjects of his lectures as well as explanations of the apparatuses he used. There were one hundred experiments for students to perform at home, along with follow-up questions and a glossary of chemical terms. The glossary was designed to aid listeners at his public lectures; the book's size (5.5 by 3.5 inches) enabled it to fit easily into a pocket.[81]

The Library Company of Philadelphia's copy of *Grammar of Chemistry* has the signature of "Martha Newbold, Philadelphia" inside the front cover. Newbold may well have had the book with her as she sat and listened to Tucker and viewed the "brilliant experiments." Or she may, like Elizabeth

Drinker, have preferred to read at home and perform her own trials. In 1802 Drinker, in her late sixties, read Adam Walker's *A System of Familiar Philosophy* (London, 1799). On June 21 she wrote in her diary: "I tried the experiment, this evening, of rubbing two peices [*sic*] of loaf sugar together in the dark, and plainly saw a luminous appearance on rubbing." On December 3 that same year she recorded: "made this forenoon a Chymical preperation [*sic*]: in a six ounce vial I put sugar of lead, fill'd it up with spring water, and suspended there in a piece of Zink, in order to produce a leaden-tree."[82] Neither sugar of lead nor zinc were common household items. Drinker must have visited the local apothecary to obtain the material she used.

Another woman, Rachel Van Dyke, was seventeen and living at home with her parents in New Brunswick, New Jersey, when in 1810 she began to read Jane Marcet's *Conversations on Chemistry*. She recorded her progress through the text and the experiments she performed with her older brother Augustus's assistance. On June 14 she wrote,

> Besides my Latin I spent nearly two hours at my Chemistry—one hour with Augustus. I think I shall like it more and more. I never expect to be a complete Chemist, but if I understand and remember what is in my book which is of the most simple sort, I shall be satisfied. I was much pleased with the experiment Augustus made this morning to convince me of the Attraction of Composition. Tomorrow we shall have another one which I long to see that is dissolving copper by nitric acid, which forms a substance entirely different from either of them in the form of little blue crystals.[83]

Two weeks later Van Dyke was already halfway through Marcet's text. She wrote of the enjoyment and excitement she experienced: "Oh! I expect I shall be a rare Chemist in time. It is not *im*probable that I may make some grand discovery. The philosopher's stone for instance—Ah—If I possessed the art of making gold—how amiable—how sensible—how all-accomplished I would be."[84] When Van Dyke completed *Conversations on Chymistry*, she was "really sorry when I came to the end of it. I was delighted with the last chapter or two. I must try to get another on the same subject but I don't believe I shall find one so interesting."[85] Six months later she borrowed the book again and read it through a second time before going on to study botany.

Science and Sociability

This private intellectual world also afforded elite women the opportunity to view scientific instruments. In Newark, New Jersey, Esther Edwards Burr boasted to Sarah Prince that she and her spouse had "a very fine *Microscope*, and *Telescope*. Indeed we have Two microscopes—and we make great discoveries 'tho not yet any new ones that I know of. The Microscope Magnifies a Lous [*sic*] to be 8 feet long upon the wall."[86]

Another purchaser was APS member James Bringhurst. Elizabeth Sandwith, the year prior to her marriage to Henry Drinker, recorded a visit to

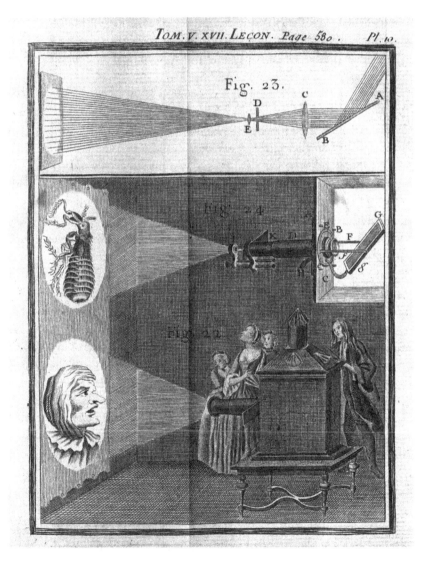

FIGURE 1.3.   Jean-Antoine Nollet, *Leçons de physique expérimentale*, vol. 5, plate 10 (Paris: H. H. Guerin & L. F. Delatour, 1777). Library Company of Philadelphia.

the Bringhursts for scientific amusement: "Spent this Afternoon, with Molly Foulk at the Widdow Bringhursts, where we were entertain'd with divers objects in a Micrescope; and with several experiments in Electricity." Thirteen years later, Drinker had not lost her interest in science. She once more recorded a visit to the Bringhursts, this time with her husband and three of their young children, where they were "entertained with sundry Electrical experiments perform'd by Js . B."[87] Science and sociability often went hand in hand. James Lackington recalled meeting his friends at the Charleston Library Society "for the purpose of improvement in science." Equipment for an evening of learning included "globes, telescopes, microscopes, electrical machines, air pumps, air guns, a good bottle of wine, and other philosophical instruments."[88]

Some colonials received instruments directly from Britain (while in London, Franklin oversaw orders of instruments for John Winthrop at Harvard College and individual purchasers such as John Bartram and Isaac Norris in Philadelphia).[89] Smaller instruments were available locally: in Philadelphia, John Breintnal advertised at his shop in Chestnut Street "A Choice Parcel of the best Spectacles, Microscopes, large and small Pocket Compasses with Dials, and several other sorts of Goods."[90] The Savannah merchants Johnson and Wylie advertised a pocket microscope along with the cloth, tools, pots and pans, tea and chocolate they imported from London and Bristol.[91] Such equipment, though not scarce, was expensive: the telescope Winthrop purchased for Harvard in 1769 cost one hundred pounds.[92] Nevertheless, there is considerable evidence that colonists purchased scientific instruments for their private use. When Sir Charles Henry Frankland, of Boston, prepared to take up his post as consul general for Portugal in 1757, he packed two of his most valued belongings—his Chelsea porcelain and his microscope.[93] The probate inventory of Nicholas Flood of Virginia recorded a "Solar Microscope and Book" along with Fontenelle's *Plurality of Worlds* and *Newtonian System of Philosophy, Adapted to the Capacities of Young Gentlemen and Ladies*.[94] Joseph Pemberton of Maryland and Peter Wagener of Virginia both owned a pocket microscope. Wagener also owned Newton's works in six volumes.[95] Doctor John Stewart of Bladensburg, Maryland, owned a set of microscopes, perhaps not unusual for a physician, but he also possessed a "Pocket Laboratory for Mineralogy."[96] The Jamaican planter Thomas Thistlewood owned two telescopes and a microscope. George Johnston, John West Jr., and Hannah Washington, all of Fairfax County, Virginia, and Samuel Hayward, another Jamaican planter, owned electrical machines. Hannah Washington's probate inventory labeled her equipment as "Priestley's machine," a reference to Joseph Priestley, who published *A Familiar Introduction to the Study of*

Electricity (1768) and sold the electrical machine described in his book.[97] Benjamin Banneker's neighbor, George Ellicott, owned several telescopes, with which, according to his daughter, "he was in the habit of giving gratuitous lessons on astronomy to any of the inhabitants of the village who wished to hear him."[98]

Despite the enthusiasm with which some men and women purchased instruments to explore nature, newspaper advertisements for the resale of instruments, especially microscopes, indicates diminishing enthusiasm for natural philosophy on the part of the owners.[99] One visitor described the state of disuse of the instruments owned by the library of Columbia, South Carolina: "The philosophical apparatus has never been in full force, except the air pump and electric machine, which were originally kept only for the inspection of the ladies, most of whom in the place have gone so far into natural philosophy as to say they have been electrified. . . . The orrery is rather out of fix, most of the planets being tired of travel and become fixed stars; they are indeed a melancholy group."[100] But if private entertainment waned quickly, public lectures, especially with demonstrations, captured the attention of Americans from the colonial era into the early nineteenth century.

James Bringhurst may have learned his parlor tricks from one of the numerous itinerant lecturers who traveled the colonies in the eighteenth century demonstrating the latest instruments and principles. These public lectures offered another venue for scientific inquiry. Adam Spencer, William Johnson, Isaac Greenwood, and David Mason traversed the Atlantic Seaboard entertaining and instructing their audiences in the wonders of natural phenomenon. Lecturers encouraged women's participation. Ann Manigault, for example, attended William Johnson's electrical demonstrations in Charleston, South Carolina, in 1765.[101] In New York, Lewis Evans offered gentlemen who subscribed for an entire course of lectures "to receive a gratis Ticket for on[e] Lady to attend the whole Course."[102] Although Ebenezer Kinnersley, perhaps the most well-traveled lecturer (he addressed audiences in Boston, New York, Philadelphia, Newport, and Virginia), was not so generous as Baron, he did offer a discount for any ladies who attended.[103]

Spectators were treated to a great deal of showmanship mixed with their instruction. Isaac Greenwood advertised that he presented his compliments "particularly to the Ladies, and Youth, of both Sexes, and assures them that no *Shocks* (as they are called) will be given on any Account." Nevertheless, Greenwood's demonstrations included "Electrified Money, which scarce any Body will take when offer'd to them," "An artificial Spider, animated by the electric Fire, so as to act like a live one," "Eight musical Bells rung by an electrified Phial of Water," and perhaps the most interesting of all, "The

Salute repulsed by the Ladies Fire; or Fire darting from a Ladies Lips, so that she may defy any Person to salute her." In an unusual gesture of inclusion, Greenwood's broadside stated: "The 5th Night he proposes to give the Black People an Opportunity of being somewhat enlightened in that pleasing noble Branch of Philosophy; which will close his Performance."[104] Whether these men and women of color were enslaved or free, Greenwood's public invitation to nonwhite Americans was unique. Certainly many of them, like Benjamin Banneker, were just as curious, and perhaps just as knowledgeable, as the rest of Greenwood's audiences.[105] Public lectures remained popular in the postcolonial era. Complemented by the increasing availability of books and periodicals and new educational opportunities for women, these informal modes of learning fostered American enthusiasm for science.

Mammoth Nation

If chemistry and electricity helped Americans understand the essence of nature, tangible interactions with plants and animals emphasized what was uniquely American about the world they inhabited. In the seventeenth and eighteenth centuries, British colonial bioprospectors scoured fields, forests, and swamps for indigenous flora and fauna. Seeds, plants, rodents, and snakes crossed the Atlantic to take up residence in the collections of the Royal Society or to root themselves in English gardens. Stones, bones, and desiccated reptiles were displayed in the cabinets of curiosities kept by colonial gentlemen and early scientific societies. A few collections contained bits and pieces of a creature referred to as the American Incognitum. Collectors were interested in the commercial viability of their finds, but they also noted which plants and animals were only found in North America. In the postrevolutionary era, fascination with the wonders of the natural world became part of the project of national identity. As Christopher Looby suggests, Americans "engage[d] in taxonomic construction as a rehearsal . . . of social and political construction."[106] Taxonomy as politics spurred Americans to inventory the natural history of the young nation. This project engendered new public entertainments in which individuals and organizations packaged spectacle, science, and political identity to create the nation's first public museums.

The first of these new sites was Pierre Eugène du Simitière's American Museum, established in Philadelphia in 1784. Daniel Bowen's Columbian Museum offered Bostonians an opportunity to view "a large collection of natural and artificial curiosities." In New York, a new political group, the Tammany Society, established their own American Museum in 1791. It was designed, as its charter stated, to collect and exhibit "everything relating to the history of America, likewise, every American production of nature or art." The city of New York, always ready to encourage "patriotic undertakings," allowed the American Museum to display its collection in a room at city hall.[107] In keeping with the museum's mission to display American products, "an Air-Gun, made in this city by an American artist," was on view. This example of American technical ingenuity fired twenty bullets in succession without recharging.[108] In Philadelphia, Charles Willson Peale also connected "rational pleasure and the instruction of the public," with national endeavors; Peale articulated the importance of studying the flora and fauna of the new nation by connecting natural history and the public good. He argued that natural history "ought to become a NATIONAL CONCERN, since it is a NATIONAL GOOD. . . . The very sinews of government are made strong by a diffused knowledge of [natural history]."[109] The exhibits of birds, mammals, insects, and rocks and minerals were displayed like an "open book of nature."[110] Peale used this conceit on the museum's tickets, which illustrated a book, opened flat, with the word "Nature" written across the pages.

Peale allowed children into the museum for free in exchange for any interesting find or gift. As a consequence of this generosity, he accumulated a large collection of hair balls from cows. Though he did not display these mundane items, the outpouring of natural curiosities illustrates the popular interest in science and nature. Museums encouraged patrons to link ideology

FIGURE 1.4. Admission ticket to Peale's Philadelphia Museum. Peale-Sellers Family Collection, American Philosophical Society.

with enlightenment in a concrete way. They could see, touch (despite admonishments not to), and smell (in improperly cured specimens) Americana.[111]

An interest in the natural products of the American continent was more than just curiosity for citizens of the United States. Americans required coal, timber, minerals, and other natural resources to develop domestic industries to replace British imports. The new nation also needed people—to produce homegrown goods, to settle new areas, and to consume American products. Emigrants were the key to success in the early republic. But generations of Britons and Europeans learned natural history with a bias against the New World—its people, its plants, its animals, and its climate. Refuting these assumptions took on a new urgency after the Revolution: national prestige and national prosperity were at stake.

No one tried harder to counter assumptions about the inferiority of American nature than Thomas Jefferson. He directed his energies where he was most likely to change people's minds. His target was Georges-Louis Leclerc, Count Buffon, the author of the thirty-six-volume natural history encyclopedia, *Histoire naturelle, générale et particulière*. Jefferson said that he took Buffon "for my ground work, because I think him the best informed of any Naturalist who has ever written."[112] Comprehensive and scholarly, Buffon's work was a damning criticism of New World animals and their environment. Jefferson claimed that Buffon "has carried this new theory of the tendency of nature to belittle her productions on this side the Atlantic."[113] In short, Buffon espoused a theory of degeneracy: animals in the Americas were weaker, smaller, and, in the case of humans, inferior to their European counterparts. Moreover, Buffon's theory about the natural environment of the Americas—damp and cold—affected even those humans and animals who migrated there from the healthier, warmer, and dryer Europe. None of this was welcome news to Americans who sought to bolster the image of the new nation. Many American reviewers of *Histoire naturelle* were dismayed by Buffon's claims that North American flora and fauna were inferior to those of Europe. The *State Gazette of South Carolina* called Buffon's description of North America a "humiliating picture indeed."[114]

Jefferson set out to convince Buffon that the Americas contained animals that were just as big and powerful as any found in Europe. He was convinced that Buffon arrived at his theory because he lacked sufficient evidence to the contrary: "It does not appear that Messrs. De Buffon and D'Aubenton have measured, weighed, or seen those of America."[115] The bones of what Americans called the Incognitum had been uncovered in Kentucky in 1739. In the 1780s, Jefferson shipped several bones to France, in the hope that Buffon would recognize them as those of the woolly mammoth, an extinct quadruped excavated in Siberia. But Buffon (correctly) identified the bones as those from another branch of the *Elephantidae* family, the mastodon. Moreover, Buffon told Jefferson that the mastodon, like the mammoth, was almost certainly extinct. Undaunted, Jefferson's next shipment to France was a large (preserved) moose. This recently living quadruped was irrefutable evidence that Buffon's theory of degeneracy was wrong. Jefferson pinned his hopes on Buffon's acknowledgment of his error in the next edition of *Histoire naturelle*. Unfortunately for Jefferson, and for American prestige, Buffon died six months after he received the moose. No revised edition appeared.

Jefferson was not alone in his endeavor to rehabilitate the nation's natural reputation. Alexander Hamilton worried that Buffon's degeneracy theory would have economic repercussions for the United States. It was important to win the psychological battle to convince Europeans (and Americans) of the nation's potential. In *Federalist No. 11* Hamilton condemned Europe's vision of itself as "the Mistress of the World." Moreover, Europe's "profound philosophers" claimed that "all animals, and with them the human species, degenerate in America—that even dogs cease to bark after having breathed awhile in our atmosphere. Facts have too long supported these arrogant pretensions of the Europeans. It belongs to us to vindicate the honor of the human race, and to teach that assuming brother, moderation."[116] Hamilton's solution was national union through a constitution: the nation's economic strength, bolstered by robust shipping networks and a navy to protect them, would refute any and all claims to American degeneracy. At the same time, selections from Jefferson's *Notes* in American newspapers and magazines acquainted readers with the mastodon. Jefferson's table of weights and measurements demonstrated that the continent held animals just as enormous as those in Europe.[117]

Hamilton's rhetoric and Jefferson's measurements were bolstered by the discovery of an almost complete mastodon skeleton in 1801. News of Charles Willson Peale's excavation of this mammoth (as Americans continued to call it despite Buffon's identification) near Newburgh, New York, fed the American fascination with all things curious, unique, and large. Peale, in the interest of self-promotion, contributed to the magnification of his find by announcing the creature as "the ninth wonder of the world." After reading Peale's proclamation, Elizabeth Drinker dryly mused, "I don't recollect hearing of the Eighth wonder, was it Genl. Washington, or Tom Paine?"[118] The excavation drew large crowds who stood for hours to watch nothing more exciting than water pumped out of a large hole in the ground. But the tedious work of excavation was forgotten once the articulated skeleton was displayed in Peale's museum, where the monster delighted and alarmed visitors. An entire room in Peale's museum, designated the Mammoth Room, was devoted to the creature.[119] After Deborah Logan viewed the mammoth, she wrote to her son, "I looked on its enormous remains with astonishment, the bones of the Head were not compleat and are supplied with wood of such a configuration as naturalists think appertains to the rest of its figure, by means of a wire the jaw opens, and displays such an extent that it frightens the Ladies, Mrs. Smyth told me of one that went to Bed after she returned home from seeing it with the terror it inspired."[120] Six years later, Peale painted an interior view of the museum, complete with the mastodon and a female observer apparently as terrorized as Mrs. Smyth's friend.

Word of this discovery fired Americans' imagination. The word *mammoth* entered common speech. The papers were full of wonders: mammoth fruit and vegetables, a mammoth pie, a mammoth steer and a mammoth calf. Suddenly, everything in the United States was big. Stirred by news of this monster of the natural world, the townsfolk of Cheshire, Massachusetts, embarked on a project of mammoth proportions: the creation of a twelve-hundred-pound, thirteen-feet-in-diameter cheese for Thomas Jefferson, the first Democratic-Republican president of the United States. Inscribed on the rind of the cheese was the motto "Rebellion to tyrants is obedience to God."[121] The cheese was the brainchild of Cheshire's Baptist minister and confirmed Democratic Republican, John Leland. Its size was dictated by the diameter of a local cider press. A special barrel was made to contain curds that were pressed, aged, and eventually turned into cheddar cheese. Leland recruited the town's dairy farmers to contribute milk for the women of Cheshire to make into curds. The scope of this collective enterprise, according to agricultural historians, required milk from the entire population of dairy cows in the town (about one thousand)—despite claims that no Federalist cows were used.

While the cheese cured, word spread of Cheshire's endeavor. Hyperbole matched the magnitude of the farmers' exertions. The *Connecticut Courant* printed an effort titled "The Mammoth Cheese. An Epico-Lyrico Ballad" that detailed the creation of the "quivering curd" and the subsequent purging of its "gushing whey" as the cider press compacted the cheesy mass.[122] In Northampton, Massachusetts, the *Hampshire Gazette* celebrated this cheese of biblical proportions with an Old Testament–style account of the undertaking:

And Jacknips [as John Leland sometimes signed himself] said unto the Cheshireites, behold the Lord hath put in a Ruler over us, who is a man after our own hearts.

Now, let us gather together our Curd, and carry it into the valley of Elisha, unto his wine press, and there make a Great Cheese, that we may offer a thank offering unto that great man. . . .

And Jacknips said, it shall come to pass, when your children shall say unto you, what mean you by this great cheese, that ye shall answer them saying,

It is a sacrifice unto our great ruler, because he giveith gifts unto the jacobites, and takes them from the federalists.

And Jacknips said, peradventure within two years, I shall present this great cheese as a thank offering unto our great ruler—and all the Cheshireites shall say Amen.[123]

While some celebrated the cheese, others ridiculed it. One Federalist newspaper stated that the Jacobites of Cheshire were mere copycats mimicking an earlier English cheese tribute given to George III. This information allowed the Federalist press to make an explicit comparison between the British monarch and Thomas Jefferson. "The Demo' Mammo' Cheese. An Epic Poem" by Don Federo, Esquire, characterized "King" Jefferson as a power- and cheese-hungry despot.[124] With Jefferson's appetite (political or otherwise) in mind, another Federalist writer suggested that the ladies of Lenox, Massachusetts, twenty miles distant from Cheshire, should bake a "great mammoth apple pye" for "King Jefferson" to eat with his mammoth cheese. The author, who signed himself "Republicanus," cautioned that the pie might require a military escort on its journey south, "lest the Federalists who our worthy President has deprived of the 'loaves and fishes,' may so hunger after his apple pye and cheese as to make an attack upon them."[125] News of the mammoth cheese also sparked reportage of other gigantic productions from the natural world—squash, peaches, and a nineteen-pound gourd. These vegetables had no discernible political affiliations, but a mammoth "good republican beet," weighing five pounds, clearly did. Its owner volunteered to turn his beet into pickles to accompany the cheese for the president's dining pleasure.[126]

In late December the cheese made its way, unmolested, to Washington. The press avidly reported on the journey. One headline blared "Mammoth Cheese Afloat!" The cheese was taken by sled from western Massachusetts to Boston. From there it was shipped down the coast to the Chesapeake Bay and into the Potomac River. The cheese made several stops along the way: the *New-York Gazette* alerted readers to the opportunity to view the curdy mammoth at a pier on the East River.[127] The cheese finally arrived in Washington on December 29. Accompanied by fanfare (and speeches), it was formally delivered to the president on New Year's Day, 1802. Jefferson was not loath to exploit this ripe opportunity for public relations. He displayed the cheese in a White House reception room. In emulation of Peale's museum exhibit, Jefferson renamed the space the "Mammoth Room." There, he invited members of Congress to view what one visitor referred to as the "New England Mammoth."[128]

Once it was ensconced in the president's house, reports of the cheese diminished. But Federalists took a parting shot when the *New York Evening Post* reported on a plan for glassmakers in the city to blow a giant bottle to hold the city's finest beer. Meanwhile, a group of bakers were called on to produce a mammoth loaf of bread, "so that [Jefferson's] convivial friends, Gallatin, Duane, and others of the same stamp, may not only have cheese, but bread, cheese and porter."[129] The same article claimed that a group of Maryland tobacco growers commissioned a pipe, with an eleven-foot-long stem, to hold a mammoth quantity of tobacco. Such an unwieldy device required eleven slaves to hold it to Jefferson's mouth.[130] Federalists employed whatever material came to hand as ammunition in their war against Democrat Republicans generally, and Jefferson specifically. But the cheese event illustrates a particularly American way of doing things in the early republic: fueled by the desire to create a national identity, Federalists and Democratic Republicans alike drew inspiration from the natural world, whose immensity seemed to confirm the importance and greatness of the American republic. The townsfolk of Cheshire, Massachusetts, in their determination to create something worthy of a president, *had* to make it big. And thanks to Peale's discovery, they had a word for outsize Americanness: mammoth.

Popular interest in the excavated mammoth as well as in the mammoth cheese was fostered by the connection between discoveries in natural history and Americans' familiarity with, and participation in, scientific ideas and innovations. It complemented the national agenda to define, promote, and celebrate American products. Jefferson used his presidency to further his desire for knowledge of the natural history of the United States. He persuaded Congress to fund the Lewis and Clark expedition in 1803, and the Freeman-Custis expedition in 1806, in order to explore and document the nation's newly acquired western lands. The aptly named Corps of Discovery returned with specimens—birds, fish, plants, and animals—and with topographic information about the previously unknown territory. Meriwether Lewis's acquisitions were displayed at Peale's museum. A Corps of Discovery member Patrick Gass rushed his journal of the expedition into print within a year of his return from the Louisiana Territory. Thus, Americans read of the expedition's journey and, in Peale's museum, they viewed the previously unknown animals that inhabited the western United States. This drive for knowledge was fed by a passion for utility. Americans looked to the natural world for resources, and to the talents of American men and women to develop products and technologies. Peale's visual record of his excavation of the mastodon is a case in point: the bones of the mastodon are not front and center in his painting. Rather, it is the cleverly devised machinery for excavating the creature that takes pride of place.[131]

Americans planted, built, dug, invented, and discovered. Lecture halls, museums, taverns, and private parlors made scientific information accessible to the curious of every age and gender. Almanacs enabled even more men and women, including people of color, to acquaint themselves with discoveries and technologies. Long before independence, North Americans were already a scientific people. Public promotion of innovation was often cloaked in a rhetoric of British nationalism—what was good for colonials would benefit Britain. As a demonstration of this belief, Benjamin Franklin presented Queen Charlotte with Pennsylvania silk, a "valuable article

FIGURE 1.5. Charles Willson Peale, *The Exhumation of the Mastodon* (1806), Maryland Historical Society.

from our colonies," as proof that Americans had succeeded in an economic enterprise that "the British Parliament seems to have had much at Heart."[132] Seven years later, in 1779, Franklin's daughter Sarah Bache requested that her father give silk to a different queen—Marie Antoinette of France—as a demonstration of what an independent America was capable of.[133] In the early United States, Americans used the experiences and lessons from their colonial past to shape a different, and according to some, unique national identity.

CHAPTER 2

Flights of Imagination

Air Balloons and National Ambitions

Aeronautics was the preeminent technology of the late eighteenth century. In the 1780s, Europeans were enthralled by the first successful attempts at flight. The Montgolfier brothers launched air balloons in Paris, and Pierre Blanchard and John Jeffries crossed the English Channel by balloon. Aeronautics was not the first technology to receive popular attention; enthusiasm for new discoveries in electricity and in the miniature world revealed by the microscope fostered numerous lectures and demonstrations. But aerial technology attracted large audiences because balloons were exhibited out of doors, not in lecture halls. Air balloons generated an excitement that spilled over into literature, consumer goods, and even medicine. After witnessing Pierre Blanchard's ascension in the prison yard in Philadelphia in 1793, the physician Benjamin Rush considered a balloon cure: "I have destined them [balloons] to be employed as remedies where a sudden and extensive change in the body is required."[1] Balloons heralded new possibilities; the world would be a different place if man could fly. Like the rest of the world, Americans were caught up in the excitement about this new technology. At first, men and women in the United States could only admire from afar. They received eyewitness reports of balloon launches in France and England. Soon, they were trying it themselves, first with miniature paper balloons, and then with full-size versions. Once Americans had mastered aeronautics, they put a uniquely American stamp on it.

Men and women appropriated balloon technology for their own practical and ideological purposes, investing it with aspirations for national prosperity: air transportation would assist farmers and merchants, and help settle the West. Air balloons came to symbolize the promise of American development. But the technology aided the darker side of national expression, too. As engines of war, air balloons helped defeat the Confederacy. As mascots of imperialism, balloons emphasized American sovereignty over conquered Indigenous peoples; balloons demonstrated racial as well as technological superiority.

Americans Witness Flight

Americans abroad were among the earliest observers of balloon ascensions. Benjamin Franklin, while living in a suburb of Paris with his grandson Benjamin Franklin Bache, witnessed several ascensions in and around the capital. In March 1784, Bache recorded in his diary, "I went to the 'Champs de Mars' to see an experiment, made by a man named Blanchard who proposed to guide an air Balloon; . . . It fell quite near its place of departure at one o'clock."[2] Three months later, he sighted another balloon: "Today at 4:45 a balloon was sent off from Versailles filled with air rarified by great heat, under the direction of Mr. Pilâtre de Rozier and a man named Proust, and it fell at 5:30 between Champ treus and Chantilly. Thus in 45 minutes it has traveled over 12 leagues. I saw it passing over Passy at 5 o'clock, disappearing at 5:10 in a cloud."[3] In September Bache viewed an ascension through one of Franklin's telescopes: "I went with my grandfather to the Abbe Armor's to see the Balloon of the Mssrs Roberts, brothers, which was to ascend. I arranged the telescopes, at eleven o'clock the ascension should begin; my grandfather was playing chess, and told me to notify him when I saw it mount. I heard a cannon fire at 3 minutes of 12, and one minute after I saw the Balloon rise, everybody gazing."[4]

John Adams and his family, during their residence in Paris in 1784–85, also witnessed these first ascensions. Like Bache, Adams's daughter Abigail kept a diary of her European experiences. She and her family were present at the ascension Bache viewed through his telescope.

> September 19th. To-day we went to see the balloon; it was to ascend from the garden of the Tuilleries; we had tickets at a crown a person to go in. We left our carriage outside and went in; the garden 1 had never been in before; it is very large, and in general, elegant. There were eight or ten thousand persons present. This people are

more attentive to their amusements than anything else; however, as we were upon the same errand, it is unjust to reflect upon others, whose curiosity was undoubtedly as well founded. We walked a little, took a view of the company, and approached the balloon; it was made of taffetas and in the form of an egg, if both ends were large; this is what contains the air; below it is a gallery where are the adventurers and the ballast. At eleven it was moved from the place of its standing among the trees to an open situation, and the cords, which were held by some of the greatest men in the kingdom, were cut; it mounted in the air. It was some time in sight, as they had intended making some experiments upon their machine. At six in the evening it descended at Bevre, fifty leagues from Paris. At two o clock the same day there was a storm of rain, with thunder and lightning, but they were not affected by it.[5]

Adams and Bache expressed the excitement and curiosity that many people experienced at the sight of this extraordinary technology. In January 1785, Bache listened to the firsthand account of Jeffries and Blanchard's cross-Channel voyage when Dr. John Jeffries dined with Franklin and his family, providing them a full account of his aerial voyage.[6] A month later, the Adamses were Franklin's dinner guests when Dr. Jeffries was again present. Abigail wrote: "Dr. Jeffries, the man of the day, I happened to be seated next at table. I made some inquiries respecting his late voyage aeriel; he did not seem fond of speaking of it; he said he felt no difference from his height in the air, but that the air was finer, and obliged them to breathe oftener, and that it was very cold. He has been so cavilled at in the papers, that I don't wonder at his reluctance at conversing upon the subject."[7]

Anticipation of an ascension could excite crowds to a fever pitch. Bache noted one such instance of the crowd's response to a failed launch in the Luxembourg Gardens: "The Balloon should have gone up at 4 o'clock, by means of air rarified by fire. The balloon took fire and the experiment did not succeed. The King of Sweden and several other personages of distinction waited until 3 o'clock. The people were furious and threw themselves upon the Balloon, and tore it in pieces each one carrying off a sample; some large enough to make a mattress and I believe the author would have been subjected to the same fate if they had not been escorted by a detachment of French Guards."[8] Many observed with horror the speed with which spectators turned into mobs. As Franklin's nephew, Jonathan Williams, commented, "Balloons and Aerial Travellers are now on a footing with Rareshows and Ropedancers."[9]

Americans abroad conveyed to family and friends back home the excitement and the hopes for balloon technology's ability to revolutionize transportation. Franklin lamented, somewhat tongue in cheek, that practical applications for balloons would not come soon enough to aid him: "The Progress made in the Management of it has been rapid, yet I fear it will hardy become a common Carriage in my time, tho' being the easiest of all Voitures it would be extreamly convenient to me, now that my Malady forbids the use of the old ones over a Pavement."[10] Franklin speculated that eventually balloons might serve as a deterrent to warfare: "Convincing Sovereigns of the Folly of Wars, may perhaps be one Effect of it: since it will be impracticable for the most potent of them to guard his Dominions. Five thousand Balloons capable of raising two Men each, would not cost more than Five Ships of the Line: And where is the Prince who can afford so to cover his Country with Troops for its Defense, as that Ten thousand Men defending from the Clouds, might not in many Places do an infinite deal of Mischief, before a Force could be brought together to repel them?"[11] While Franklin imagined practical balloon applications far in the future, in the here and now Abigail Adams did not believe that balloons were a safe mode of transportation. She wrote to her sister, "You talk of coming to see us in a Balloon. Why my Dear as Americans sometimes are capable of as imprudent and unadvised things as any other People perhaps, I think it but Prudent to advise you against it. There has lately a most terrible accident taken place by a Balloons taking fire in the Air in which were two Men. Both of them were killed by their fall, and there limbs exceedingly Broken. Indeed the account is dreadful. I confess I have no partiality for them in *any* way."[12] Despite her misgivings, the following year Adams playfully entertained the notion of a balloon to convey one of her nieces across the Atlantic: "I wish I Could send a Balloon for one of my Nieces. I shall want a female companion Sadly."[13]

Only a handful of Americans received correspondence from family or acquaintances in Europe, but information about balloon technology and balloon ascensions was readily available to the rest of the country in newspapers and magazines. Only months after Jacques-Étienne and Joseph-Michel Montgolfier's ascension in Paris in August 1783, the *Salem Gazette* gave readers a detailed account of the launch. This technological feat inspired the writer to suggest specific applications for this new form of transportation: "It may be applied by navigators and travelers to reconnoiter the tracts of country, and by others to reconnoiter the position of fleets and armies, and prevent surprises from them." Balloons might also "animate ambition in science, and spread a desire of knowledge." The *Gazette* cautioned that balloon technology should not be dismissed as useless. It reminded readers that "almost all great improvements having been gradual and unexpected. It would be wiser therefore to improve the machine in question, and let knowledge take its course."[14] Several months later, American papers printed Benjamin Franklin's eyewitness account of a successful ascension by the Robert brothers in December 1783. Franklin noted that "All Paris was out to see it, and all the inhabitants of the neighboring towns, so that there could hardly be less than half a million of spectators." Franklin called aeronautics "the most extraordinary discovery that this age has produced."[15]

Between February and August 1784, the *Boston Magazine* published a series of articles reporting in detail the French ascensions of the Montgolfier brothers. The magazine devoted several pages to aerostatic news, and it printed one of the first American images of an air balloon.[16] In November, the *Pennsylvania Gazette* published a translation of Pierre Blanchard's account of his third ascent in July of that year. Americans were clearly interested in and enthusiastic about this new technology.

Soon almanacs used balloon images to attract customers. One of the first to do so was *Weatherwise's Town and Country Almanac for 1785*. The cover displayed the striking image of an aerial disaster. The illustration, and an accompanying poem on the first page, reminded readers that this technology was still in the experimental stage. Such disasters might serve as a caution to Americans to shun such untested technology. The almanac warned: "READERS behold a Plate design'd / To lend a moral to Mankind." The poem likened courtiers, grasping lawyers, and greedy merchants to air balloons, all of which were "little else, but silk and gauze / Prunella, puff, and bursting Noise."[17]

Other almanac makers went a step further: new almanacs appeared with titles that blatantly pandered to Americans' enthusiasm for air balloons. John Steel's *The Balloon Almanac for the Year of Our Lord 1786* (Philadelphia, 1785) included "A Short Account of Mr. Lunardi's Aerial Voyage, from the Artillery-Ground, London" and, on the same page, Philip Freneau's poem "The Progress of Balloons."[18]

Almanac illustrations of balloon ascensions (and failures) gave most Americans their first image of a balloon, but a few people were already trying their hand at balloon construction. Francis Hopkinson of Philadelphia informed Franklin that Parisians were not the only ones who were eager to make and fly balloons: "We have been diverting ourselves with raising Paper Balloons by Means of burnt Straw to the great Astonishment of the Populace—This Discovery, like Electricity, Magnetism and many other important Phenomena, serve for Amazement at first—it's Uses and Applications

FIGURE 2.1. *Weatherwise's Town and Country Almanac for 1785* (Boston: Weeden and Barrett, 1784). American Antiquarian Society.

will hereafter unfold themselves."[19] Hopkinson provided Thomas Jefferson
with a more detailed account of Philadelphians' first attempts at flight:

> We have been amusing ourselves with raising Air Balloons made of
> Paper. The first that mounted our Atmosphere was made by Dr. Foulk
> and sent up from the Garden of the Minister of Holland, the Day
> before yesterday. Yesterday Forenoon the same Balloon was raised from
> Mr. Morris's Garden, and last Evening another was exhibited at the Min-
> ister of France's, to the great Amusement of the Spectators. They rose
> twice or perhaps three Times the Height of the Houses, and then gently
> descended, without Damage. They were open at Bottom, and of Course
> the Gas soon wasted. I am contriving a better Method of filling them.[20]

Jefferson was soon in Philadelphia to observe balloons for himself. On
May 17 he purchased tickets to watch an ascension.[21]

At Mount Vernon, George Washington read newspaper accounts of the
French aeronauts. He was eager to learn more: "I have only newspaper Accts
of the Air Balloons, to which I do not know what credence to give; as the
tales related of them are marvelous, & lead us to expect that our friends at
Paris, in a little time, will come flying thro' the air, instead [of] ploughing
the Ocean to get to America."[22] Washington cautioned his English friend
Edward Newenham against a proposed ascension: "Had I been present &
apprized of your intention of making an aerial voyage with Monsr Potain,
I should have joined my entreaties to those of Lady Newenham to have pre-
vented it." Washington believed that ballooning should be left to "young
men of science & spirit to explore the upper regions," and that such men
should be financially compensated for the risks they took.[23]

Despite Washington's caution, Virginians soon had balloons. In Richmond, a Mr. Busselot, from France, launched a balloon in the public square
in front of the capital.[24] In Williamsburg, Jefferson's friend, the Reverend
James Madison, lamented his inability to raise a balloon: "We have not yet
however been able to ascend. I find the Air from Straw much more inflammable than any I ever collected from Wood &c., but have not been able to
observe that extreme Levity, which must be necessary for the Purposes of
a Balloon."[25] Two years later, Madison wrote to Jefferson that he had succeeded. But he noted that "no one in America has yet ventured to mount
with a Balloon."[26] That soon changed.

Balloon Influenza

The first manned flight in the United States took place in Baltimore on June
24, 1784, when a young boy, Edward Warren, volunteered to go up in the

aerialist Peter Carnes's balloon.[27] Three weeks later, Carnes announced his own flight from the prison yard in Philadelphia. Carnes's successful Baltimore launch had taught him that his balloon required protection from the curious and the overly enthusiastic; his advertisement reminded citizens that only ticket holders would be admitted to the enclosure to view the balloon, where armed guards were posted around it. Carnes threatened that he was justified "in taking the life of any person who attempts to force his way into the field."[28] Unlike Carnes's success in Baltimore, the Philadelphia ascension was nearly a disaster: Carnes fell out of the balloon when it was ten feet from the ground, just before the balloon burst into flames.[29]

Despite the perils of flight, Americans embraced the technology as proof of the nation's potential. One Boston paper chided critics who dismissed the talents and ambitions of the young nation: "The opinion that philosophy will never spread her rays in the new world, can certainly have no foundations in truth."[30] A Baltimore paper pointed to Carnes, the first American to launch a balloon, as proof of American ingenuity: "'The MUSES follow FREEDOM,' said Socrates. From Greece and Rome they certainly fled when those mighty Empires fell. Let us hail therefore their Residence in America!" Recalling the martial triumph over Britain, the author added, "I am pleased in reflecting how much our Countrymen have done to improve the various Branches of Science, and doubt not our being as much distinguished for Works of Genius in Times of Peace, as our Patriot Army have been for their Success and Sufferings during the War."[31]

Most Americans did not witness a balloon ascend, let alone fly in one. Nevertheless, they demonstrated enthusiasm for the new technology. Francis Hopkinson lamented to Thomas Jefferson that the popularity of balloons eclipsed more weighty concerns: "The Name of Congress is almost forgotten and for one Person that will mention that respectable Body a hundred will talk of an Air balloon."[32] Merchants took advantage of the balloon's appeal with clever advertisements: the Boston shopkeepers J. L. and B. Austin used their customers' familiarity with air balloons to assert the quality of frying pans, hardware, corduroys, and other imported goods: "They do not recommend their Assortment as BALLOON, or INFLAMMABLE; but as Articles consisting of the DURABLE and SUBSTANTIAL."[33] This aerial enthusiasm prompted one Charleston, South Carolina, merchant to promote his Madeira as having been "imported in an AIR BALLOON."[34] Enterprising shipowners christened their vessels Air Balloon. Brigs, schooners, and sloops were emblazoned with the name.[35] One ingenious American applied for a patent for a "Federal Balloon." This invention was not a balloon at all, but some sort of machine for exercise. He clearly meant to capitalize on public enthusiasm to sell his device.[36]

Apparel embellished with balloons filled the shops. There were balloon handkerchiefs and scarves, balloon fabric, balloon hats, buttons, fans, bonnets, stockings, and ribbons.[37] Gentlemen purchased Air Balloon umbrellas and Air Balloon watch chains.[38] Newspapers reported with amusement the enthusiasm with which American women embraced balloon fashions. One commentator urged women to increase the size of the hoops at their hips so as not to be carried off in the air by their balloon hats.[39] Under the headline "Fashion," a New Jersey paper informed its readers that "the balloon-hat, after having been superseded for a few days, by the adoption of the rural straw umbrellas, has again been reinstated as the capital ornament of female undress."[40] One New Jersey wit composed "A Balloon Song" about fashionable balloon attire:

> The bonnet, hat, cap and toupee,
> Erected in the air must be—
> Ev'ry lovely charming she,
> Must have her Hair—Balloon O.
>
> That you may see how nice they stride,
> And where-about the stocking's ti'd,
> Each petticoat's extended wide,
> And is an Air-Balloon O.[41]

"Balloon influenza" raged in the 1780s. One commentator wrote that "nothing can pretend to have any intrinsick value in it, unless it has this *name* as an appendage . . . as a countryman was heard one day last week—'*Fine balloon string beans.*'"[42] The Boston *American Herald* chose to celebrate "balloon Madness" with a "Balloon Wish" for New Year's 1785:

> In this wild, romantic Age,
> What fantastic Whims engage!
> High and low and old and young,
> All with balloon Madness rung!
> Balloon Hats and frying Pans,
> Balloon Ribbons, balloon Fans,
> Balloon Gauzes, balloon Caps,
> Balloon Hoops, or Balloon Traps![43]

A Connecticut paper reprinted a lament from Jamaica, which, noting an absence of local news, suggested that printers solicit news items to catch the reader's attention: "Wanted, a war between the Emperor and the Dutch—a bloody battle on the Musquito Shore—a tumult among the Congressional gentry at New York—the death of some eminent man—a horse race— . . . a runaway match between a pair of fond lovers, or any kind of intrigue—a balloon—a duel—a column or two of scandal."[44]

Balloons were celebrated in verse and prose. The poet Philip Freneau, among the first Americans to imagine the possibilities of air travel by balloon, was enthusiastic about practical uses for the technology:

> The stagemen, whose gallopers scarce have the power
> Through the dirt to convey you ten miles an hour,
> When advanc'd to balloons shall so furiously drive
> You'll hardly know whether you're dead or alive,
>
> Yet more with its fitness for commerce I'm struck—
> What loads of tobacco shall fly from Kntuck,
> What picks of best beaver—bar iron and pig,
> What budgets of leather from Conococheague!
>
> To market the farmers shall shortly repair
> With their hogs and potatoes, wholesale, thro' the air,
> Skim over the water as light as a feather,
> Themselves, and their turkies conversing together.[45]

Other writers imagined balloon travel among the stars and planets. William Woodhouse advertised his six-part poem, "Balloon, or an Aerial Trip to Celestial Regions."[46] A novel, *An Account of Count D'Artois and his Friend's Passage to the Moon In a Flying Machine, called, An Air Balloon*, related the adventures of D'Artois and his friend and companion, Vogrill. They voyaged to the moon, met its inhabitants (who conveniently speak Hebrew), and conversed on war, politics, and society.[47] Even foreign works were co-opted by Americans: when the British playwright Elizabeth Inchbald's farce about an imagined balloon trip, *A Mogul Tale*, was performed in New York, the theater gave the play a local setting. Rather than an excursion to the Middle East, the characters land in New York City. Balloons were seen as an American enterprise, even in moments of folly.[48]

The rhetoric generated by balloon launches makes clear that elites felt a proprietary interest in science-related events. Reports often described the "genteel" or "polite" observers of an ascension. Satirical fiction and verse implicitly delimited the appropriate audience for (and consumers of) the new technology. In a letter addressed "To the Air-Balloon Maker," the author complained that, as a result of the unnamed aeronaut's announcement of an intended balloon launch, the author's wife "fell in such love with the aerial excursions that nothing less would satisfy her curiosity than that she would build a balloon at our own expense, and make an experiment." The indulgent husband declared that the 150 pounds his wife spent on the balloon was nothing when compared to the "gratification of her fancy." The author, his wife, and their two children all ascended. By misadventure, the author tumbled out, and watched with horror as his wife and children drifted away in the sky. In a parody of the many runaway advertisements in American newspapers, the distraught husband requested that all newspapers carry the following notice: "On the 12th instant left this lower world, in company with a little boy of 3 ½ years old, and girl of 2, *Mary Harlow*, aged 31 years, of middle size; had on and took with her, 1 green silk dishabille, one check muslin ditto, a blue riding dress, her hair hanging in gentle ringlets on her shoulders, walks very stately, speaks Dutch and French, and a doggrel of Latin.—Whoever will conduct the lady to the bearer, or give information to me, Jonathan Harlow, at the Tea water pump, in what region she has taken up her abode, shall be handsomely rewarded."[49] Women were belittled for a superficial fascination with balloons; consumption of balloon hats, buttons, umbrellas, and stockings proved that female engagement with the technology was driven by fashion and curiosity, not by scientific inquiry. Female attempts to engage with scientific learning and technologies, as Mary Harlow's experience proved, could only end in disaster.

Race as well as gender occasionally figured in fictional reports of balloon ascensions. A correspondent to the *New York Journal* wrote of his delight in observing "the variety of emotions, which the passing of the balloon, . . . occasioned among the several classes of people." To "amuse your readers," the author included an imagined vignette: "Sally, Polly, Dick, Sambo, run quick (says the good lady to her family) to the chamber window, to the top of the house!—Wat de matter misse (says Sambo) The balloon, you booby, is right over your head—Yi—Yi—Yi— (says Sambo) with a shriek that might have been heard to Bowery-lane, and away he went down the arch-cellar, followed by Sally, Polly, and Dick, like lightening, for fear the balloon would break their skulls, and left the good old lady to gaze alone."[50] The article presents racial and gendered hierarchies from the presumably male omnipotent narrator to the marveling female and her uncomprehending slaves. Balloons were proprietary: only elite white males could properly understand, appreciate, and use them.

The enthusiasms of white women and the apprehensions of enslaved Black men were dismissed as misguided and superficial (in the case of women) and as superstitious (in the case of slaves). But balloon technology received serious inquiry from scientific organizations such as the American Philosophical Society. Inspired by European achievements and, closer to home, by

Peter Carnes's fledgling attempts at flight, a group of "respectable citizens of Philadelphia," including several American Philosophical Society members, invited the public to subscribe to the development of an air balloon. Their appeal for funding made it clear that such a project was important to the national project of improvement. "To the Citizens of Philadelphia" first appeared in Philadelphia newspapers five days after Edward Warren became the first person in America to ascend in a balloon. The names of more than sixty Philadelphia citizens of the first rank, including the mayor, Samuel Wharton, the physicians Benjamin Rush and John Morgan, the artist Charles Willson Peale, John Vaughan, John Swanwick, and unnamed "Professors at the University" were listed as subscribers. The group assured readers that their project was not for mere amusement. It had the highest motives: "the joint aid of the enlightened and patriotic sons and daughters of American freedom and science." Such an endeavor would "exalt our national character for philosophy and love of science."[51] The group appealed to the practical concerns of Americans; with sufficient funding and American ingenuity, balloons would end wars because "quick advices may be given of Intended invasions, which may be thereby rendered abortive." Balloons would also increase commerce: "Discoveries of new, or a more thorough knowledge of back countries may be made, trade will be improved." And further research into aerostatics would benefit other branches of science, such as astronomy and electricity. The broadside emphasized that it would be Americans who made these further discoveries and improvements. Americans had already proven their abilities: Godfrey, who invented the quadrant, and Franklin, who invented electric rods, "for the honor of our country, were born in America." There was nothing balloon technology was not capable of improving in the new nation. One wit suggested that balloons could be put to use for the benefit of the government: congressmen might "float along from one end of the continent to the other, observe that the systems of government are properly supported, and when occasion requires can suddenly pop down into any of the states they please."[52]

In light of the excitement about the first American balloons, Pierre Blanchard's ascension in Philadelphia in January 1793 might appear anticlimactic. Public enthusiasm had already reached fever pitch; Peter Carnes had made history. What did Blanchard offer Americans that they did not already possess? In 1785 John Quincy Adams commented, "All that has as yet been done relative to this discovery, is the work of the French. Montgolfier, Pilâtre de Rozier, and Blanchard will go down, hand in hand to Posterity."[53] Now the illustrious Blanchard came to America. Not only did Blanchard's ascent receive nationwide coverage and fanfare; it was sanctioned by the

United States government. The highest federal officers, including President Washington and Secretary of State Thomas Jefferson, witnessed Blanchard's launch.[54] Benjamin Franklin Bache, now editor of the Philadelphia Democratic Republican newspaper the *Aurora*, once again witnessed a Frenchman ascending in a balloon—this time on American soil. Blanchard waved the US flag as he rose from the ground. President Washington removed his hat and bowed to the aeronaut.[55] For the physician Benjamin Rush, the sight of Blanchard's balloon ascending into the sky was "truly sublime."[56]

Blanchard's ascent was witnessed by thousands. Samuel Mickle, of Woodbury, New Jersey, canceled his planned trip to Philadelphia when his wife reminded him that Blanchard's launch was to take place there. Mickle wrote in his diary, "all Woodbury almost was going to see it which appearing likely to obstruct my business with some people there have postponed going there." After Blanchard came down within a mile of Mickle's house, he noted the "Great ado with looking for and at the Balloon. . . . Balloon is ye subject in almost every quarter."[57] Benjamin Rush reported that spectators had come from "New York, Baltimore, and other distant parts to see it. The city was so crowded that it was difficult for strangers to get lodgings at taverns, and the theatre was so crowded this evening that several hundred people returned without getting in."[58] Blanchard's well-advertised launch attracted spectators who wished to see the famous French aeronaut. But American commentators placed the Frenchman into an American context. As one article proclaimed, "Drawing lightening from the clouds was an experiment that must immortalize the fame of this country with the name of the man who first conceived the idea—visiting those clouds is an experiment worthy of patronage from the country that gave birth to the first." The shadow of Benjamin Franklin, an American, towered over Blanchard, a Frenchman.[59] Blanchard's launch was a way of confirming the promise and importance of flight. If ballomania affected Americans, it did so indiscriminately. Despite Jonathan Williams's fear that popular enthusiasm diminished the scientific value of aviation, the federal government's participation in Blanchard's flight signaled that the enterprise was worthy of official encouragement.

Aeronautics captured the public's attention in profound and sometimes disturbing ways. In September 1819, Philadelphia's *Franklin Gazette* announced a balloon ascension at the city's own Vauxhall Gardens. At four in the afternoon, the French aeronaut Michel would enter the basket of his hot air balloon, ascend with it, and then jump from the basket with a parachute and float back into the gardens. Five days prior to Michel's ascension in Philadelphia, another aeronaut staged a balloon launch just across the Delaware

River, in Camden. As many as twenty thousand spectators "assembled at the above place, and on the different streets, wharves, and houses, near the river." Philadelphians eagerly crossed the Delaware to observe the ascension: "Every boat was put in requisition. The ferry boats that ply between the city and Camden were filled with passengers: and so great was the press to get on board of them, that we understand a number were pushed off the slips, but were immediately taken out of the water." But the balloon failed to rise. Frustrated at their wasted journey (and wasted money), a few individuals "meanly cut slits in the balloon with their knives."[60]

Denied their view of an aerial ascension in Camden, Philadelphians awaited Michel's balloon launch at Vauxhall Gardens. The event drew hundreds of people to Vauxhall and thousands more to locations around the city where they could see the balloon without paying the one-dollar entry fee. Individuals and families came into town from the hamlets and villages surrounding Philadelphia. People gathered on hills along the Schuylkill River and on the road outside the venue. Although these observers could not watch Michel begin his ascent, they would easily see the balloon once it rose over the Vauxhall Gardens' fence. One observer, Samuel Breck, recorded in his diary that he "walked with Lucy and some other children to the high grounds of Mantua village, from which we had a good view of the garden. We waited until past 6 o'clock, without being gratified with this novel sight, and returned home."[61] Breck did not know why the balloon failed to ascend. Nor was he aware of the riot in the city beneath his gaze. Due to insufficient gas, Michel's balloon did not rise. The crowd outside the fence became restive. Then the trouble began.

An impatient few threw stones that punctured the balloon. Others attempted to climb the fence, gave up, and simply tore it down. The catalyst for the worst of the crowd's rage was the cudgeling of one of the fence climbers, a boy of fourteen, by a Vauxhall guard. The sight of the boy, bloody, unconscious, and presumed dead, stirred the outsiders to action. Using the Vauxhall flagpole as a battering ram, they broke down the gate. Once inside, the crowd stormed the now sagging balloon and tore it to pieces. Frightened musicians, who had serenaded the paying guests while they waited for the launch, ran for their lives as ravagers grabbed and smashed the instruments, and hacked a piano to pieces. The refreshment bar was also destroyed, though not until all the alcohol was consumed.[62] Like the furious Parisians who tore a balloon to shreds in 1784, these balloon-besotted Philadelphians were enraged at being denied their entertainment. As Benjamin Franklin commented to a friend after the Paris incident, "It is a serious thing to draw out from their Affairs all the Inhabitants of a great City and its Environs, and a Disappointment makes them angry."[63] The Vauxhall Gardens debacle illustrated the dark side of popular enthusiasm. But these instances of lawlessness also demonstrated how avidly the public followed the activities of early nineteenth-century aeronauts. Even though the American imagination ran far ahead of the technology, transportation by air became a permanent part of how Americans envisioned their future.

Airy Ambitions

Americans continued to hope for, and to imagine, applications for the infant technology. In the coming years, there might be, according to one newspaper, balloon coaches to carry passengers afar.[64] In the story "A Peep into Futurity," a sleeper from 1850 awakes in 1950 and marvels at the technologies that speed passengers through space—including balloons.[65] Ambitions for speedy transportation were also reflected in *Ladies Rights! Or Fitchburg 100 Years Hence*, performed in Fitchburg, Massachusetts, in 1854. In the play, balloons, or "aerial locomotives," whisked passengers to California, London, Paris, and China. Private balloon hires conveyed honeymoon couples to the Sandwich Islands.[66]

In the here and now of the early nineteenth century, Edgar Allan Poe's balloon hoax, given legitimacy by a prominent New York newspaper, demonstrates the confidence Americans placed in the new technology. In a one-page extra edition on Saturday, April 13, 1844, the *New York Sun* announced that the aeronaut Monck Mason successfully crossed the Atlantic in a balloon. The headline screamed "Atlantic Crossed in Three Days!"[67] What made Poe's fictional report so plausible, and attractive, was the fact that Mason had made a successful journey from Britain to Germany several years earlier. And the American aeronaut John Wise was on the brink of attempting an Atlantic crossing.[68] Everyone knew Mason really had succeeded in a long-distance journey by air, and Mason published a pamphlet recounting his voyage and describing in detail a model airship he had demonstrated in London seven years later. Thus, Poe had ample material to ensure that his story of Mason's Atlantic crossing sounded real.[69]

Poe's balloon hoax in the *Sun* was not his first attempt to capture the public's attention with plausibly implausible balloon news. His 1835 story, "The Unparalleled Adventure of One Hans Pfaall," recounted the experiences of a Dutchman who traveled to the moon by balloon.[70] Poe may not have contrived "Hans Pfaall" to pass as fact rather than fiction, but his depiction of Monck Mason's voyage was close enough to actual aerial accomplishments to catch the public's attention. Poe himself acknowledged that his description "was written with a minuteness and scientific ability calculated to obtain credit everywhere, and was read with great pleasure and satisfaction." Even Poe's retraction, published in the *Sun* two days later, acknowledged that "we by no means think such a project impossible."[71] Despite the fact that the *Sun* could not confirm the truth of the account, the accuracy with which the balloon was described—a description that fit closely the real balloon Mason used in his Channel crossing—suggested that a transatlantic voyage was certainly possible and perhaps imminent. Poe's elaborate leg-pull resonated with an American public already invested in balloon technology.

Part of this investment involved children. If air travel was not yet a reality, the attention paid to educating the next generation of Americans about the technology was proof that it was something they needed to understand. One of the earliest books from which children could learn about the science behind air balloons, and from which they were instructed in making their own, was *The Art of Conjuring Made Easy* (1823). These were small balloons, like the ones Francis Hopkinson amused himself with in the 1780s. Creating inflammable air to fill the balloon required oil of vitriol, iron filings, and much patience. Children needed indulgent parents for both the supplies and the assistance. Thus, early nineteenth-century science education expanded basic understanding of chemistry to a practical application—flight. By 1833 *Parley's Magazine for Children and Youth* assumed that at least some of its young readers had seen a balloon ascension. The article begins "Did any of my young readers ever see a balloon? . . . Perhaps some of you, who live in the city of New York, have witnessed the balloon ascensions of Mr. Durant, and can readily say 'yes' to my question." The article related the history of balloon experiments and early ascensions in France and England. It brought readers up to date with an account of Durant's flights: "Mr. Durant, of New York, is, I believe, at present, the chief, if not the only aeronaut, in the United States."[72] The article announced that Durant would soon make an ascension from Albany, New York, and encouraged "our young friends in that quarter [to] send us an account of his ascension." Obligingly, one reader did so. In the very next issue, "a Young Correspondent" provided his eyewitness account of Durant's aerial voyage: "At last the cords were cut, and my heart beat as if I were going up in the air myself. The people shouted, and I shouted, and everybody shouted. Mr. Durant waved a flag as he rose. The balloon rose up like a bird, and sailed away till it seemed like a speck. At last it flew out of sight; and, taking me down from his shoulders, my brother returned home with me."[73] Just like the adults, children were balloon enthusiasts.

Even very young children learned about balloons. Mary Swift's *First Lessons on Natural Philosophy for Children* answered basic questions: "How are balloons made?" "Do accidents ever happen to those who ascend in balloons?" The answer was "Very often."[74] Theodore Thinker's *The Balloon and Other Stories* also warned its young readers: "It is not very safe to travel in balloons."[75] A short story in *The Child's Gem* presented balloons and steam engines as rival technologies. In "The Balloon and the Steamboat," two boys debate which technology is superior. The debate ends when a balloon overhead bursts and the aeronaut falls into the lake, only to be rescued by a passing steamboat.[76] Despite these fictional catastrophes, stories and essays for children continued to offer information about balloon technology and to entertain readers with stories of balloon voyages. Some anthologies offered brief accounts of ascensions, such as "The Balloon" in *Aunt Mary's Stories for Children*.[77] Other publications, aimed at older children, provided more technical information. *Robert Merry's Museum* published an essay titled "Aeronautics, or Art of Navigating the Air with Balloons." It promised that, "in view of its interest, especially to the young reader, we shall devote a few pages to a notice of some of the most splendid aerial voyages which we find on record."[78] The *Museum* sought to engage children's imagination as well as instruct them about balloons. In a series of essays, the character Robert Merry took several girls and boys on a balloon voyage through Europe, allowing them a bird's-eye view of rivers, mountains, and cities and conveying geography lessons in an entertaining format.[79]

The emphasis was on learning: books and stories about air balloons groomed children to pursue studies in science and technology. A balloon story in *The Evergreen: or, Stories for Childhood and Youth* explicitly linked the excitement generated by air balloons with the imperative to study. Old Giles tells young Harry that he, too, might grow up to be a man of learning like the aeronauts: "Who knows, Master Harry, if you are not an idle young gentleman, but mind your lessons rather than spend all your time in play—who knows, I say, what wonderful thing you may one day find out?"[80] But play was learning too. An advertisement for a book titled *Parlour Magic* promised the boy, "whose wonder has been excited by the scientific lecturer, . . . fifty-six amusements in Light and Heat; forty different experiments with Gas and Steam; eighty with air and water."[81] In the 1780s, Francis Hopkinson complained to his friend Benjamin Franklin that education was misdirected: "A great deal of precious Time is spent in favoring upon young Minds logical and metaphysical Subtleties—which can never afterwards be applied to any possible Use in Life—whilst the practical Branches of Knowledge, are either, lightly glanced over, or totally neglected."[82] Fifty years later, Hopkinson's ambitions to rear a generation of scientists, inventors, and technicians had been fulfilled.

Civic Celebrations

The rhetoric that promoted air balloons as the key to better transportation, greater commerce, and well-educated children was more than simply hot air; communities chose to embellish public activities with balloons, and balloons became a mascot for civic pride. One of the earliest displays was the 1793 Boston celebration by Democratic Republicans in honor of French Revolutionary victories. The festivities included an unmanned balloon emblazoned with an American flag and the words, "Liberty and Equality."[83] By the early nineteenth century, balloons were a necessary component to a successful celebration. The "Grand Procession of the Victuallers of Philadelphia" in 1821 is one example among many. The parade was orchestrated by a Mr. White to promote his business and that of his fellow cattlemen. White's announcement of the coming parade in the *Pennsylvania Gazette*, while self-promoting, was couched in terms of civic utility; the occasion was "highly interesting and honorable to the character of the State of Pennsylvania for the improvement of the breeds of domestic animals, and equally important for its agricultural concerns."[84] Driving home his point that successful business was an ornament to the republic, White assured his readers that the parade "will be one of the most brilliant spectacles ever exhibited since the celebrated Federal Procession." *Poulson's American Daily Advertiser* applauded the "public Spirit" behind this cattle extravaganza.[85] Thousands of pounds of beef on the hoof were led down Market Street past crowds of spectators. Newspapers reported on the number and variety of animals, including sheep, pigs, and cattle along with the amounts of meat obtained after slaughter. There were four thousand pounds of sausages alone. The *Pennsylvania Gazette* noted that several air balloons ascended during the parade, a phenomenon that "added considerably to the novelty and beauty of the scene."[86] John Lewis Krimmel's painting of the enormous civic stock parade shows one of the balloons hovering overhead. Pride, private enterprise, and a rhetoric of boosterism were articulated with words, activities, and objects—all meant to wrap an otherwise mundane display of cattle with an aura of civic engagement and progress. What better way to express progress than with an air balloon?

When the Marquis de Lafayette returned to the United States in 1824, towns and cities throughout the Northeast greeted him with fanfare, celebrations, and ceremonies. Advertisements for souvenirs and keepsakes

FIGURE 2.2. John Lewis Krimmel, *Procession of the Victuallers* (1821). Acquatint by Joseph Yeager. Library Company of Philadelphia.

flooded the newspapers. One commentator noted that "Everything is Lafayette, whether it be on our heads or under our feet," the *Saturday Evening Post* commented in the aftermath. "We wrap our bodies in Lafayette coats during the day, and repose between Lafayette blankets at night."[87] One enterprising barber announced that he had devised "a new and original mode of hair cutting, neither imported from England nor France, but truly American, contributing as I would wish to all do honor and merit to our second Washington—I have denominated it 'A la mode de la Fayette.'"[88] In Philadelphia, architect William Strickland designed a triumphal arch in front of the State House. Although it was a temporary structure made of wood and covered with painted canvas, the effect was nonetheless impressive. The procession for Lafayette's entry into the city drew large crowds along the route. Lafayette's carriage drove through the arch and deposited the general at the door of the State House. Once inside, Lafayette was feted with speeches and with reunions with old soldiers from the Revolution. Outside, crowds continued to cheer, to marvel at the magnificent ceremonial arch, and to admire the air balloon launched above the city to honor the occasion. Keepsakes continued to flood the market long after Lafayette's visit, including handkerchiefs produced by the Germantown Print Works. The vividly detailed image shows the general's carriage passing through the triumphal arch while the balloon rises above the State House.[89]

If air balloons reinforced the importance of an event to a community or to the nation, reciprocally, citizens were sometimes called on to lend purpose and importance to balloon ascensions. When the aeronaut Charles Durant announced his ascension at Castle Garden in New York City in 1830, he invited the Common Council of New York to attend. They did. Durant used patriotism to whip up enthusiasm for his coming voyage. In "The Aeronaut's Address," distributed to the crowd before his flight, Durant hailed the United States as "The country of freedom and honor, / A home for the brave and opprest." A band played a selection of "national airs" as the balloon inflated. Durant waved the American flag as he passed over the spectators on his route.[90] Further cementing the bond between aerial flight and national progress, the balloon was painted with the visage of Governor De

FIGURE 2.3. Lafayette's arrival at Independence Hall, by Germantown Print Works, 1824–25, Germantown, Philadelphia. Courtesy of Winterthur Museum.

Witt Clinton—the politician most lauded as responsible for another triumph of American technology, the Erie Canal.[91]

American Power

Balloons were employed as symbols of civic pride and celebration, but they were also put to use as symbols of national power. Durant's sixth ascension in New York City in 1833 coincided with the visits by Andrew Jackson, the American president responsible for Indian removal, and Black Hawk, the Sauk leader recently defeated by the US military and Sioux allies. Black Hawk and his diminished band of Sauk and Fox were captured, briefly imprisoned, and then taken east to Baltimore, Philadelphia, and New York.[92] Durant's balloon ascension was planned to take place before both the conqueror and the vanquished. Spectators thronged to Castle Garden, encouraged by "the prospect of beholding at once the President, the Indian Chiefs, and the balloon ascent."[93] Jackson, pleading fatigue, left Castle Garden before the balloon rose. Black Hawk and his party, aboard the steamboat *New York*, did view the ascension. No mention was made of the fact that Black Hawk and his fellow Sauk were literally a captive audience—forced to watch a demonstration of American technological superiority. Lewis Cass, Jackson's secretary of war, had ordered the captives sent east on a tour of the major cities. The journey had the twofold goal of confronting the Native Americans with a civilization mightier than their own, and of showing European Americans a defeated and now subject people. As the *Richmond Enquirer* described the situation, "It is the desire of the government that the present chief may see and judge for himself the extent of the people with whom they presume to war."[94] Cass had little regard for the rights, culture, or even intelligence of Native Americans. To him, Indians were backward and uncivilized. Cass's opinion of Native Americans, shared by many, was reinforced by craniologists and phrenologists. The physician Samuel Morton's published study of the size, shape, and brain capacity of Indigenous Americans, *Crania Americana*, stated that native peoples were "averse to cultivation, and slow in acquiring knowledge; restless, revengeful, and fond of war, and wholly destitute of maritime adventure."[95] Shortly after the Native Americans' New York visit, the *American Phrenological Journal* published an essay on Black Hawk. According to measurements taken from a plaster bust, the Sauk leader's organs for destructiveness, combativeness, secretiveness, and self-esteem were "all very large." The essay informed readers that "these organs, when large, or very large, always give great energy and force of character, and, in a savage state, would give cruelty, cunning, and revenge; would make an Indian the bold and desperate warrior, and tend to raise such a one to be a leader, or chief, where physical power and bravery are the most important requisites."[96] As with most phrenological analyses, the conclusions about Black Hawk were post hoc. The fact that the Sauk, and many other tribes, vigorously fought back against American expansion made Cass more determined to bring them to a sense of their subjectivity.[97] Black Hawk was meant to see the "extensive Republic, studded with cities, towns, and prosperous farms embellished with all the improvements which art can devise or industry execute."[98] In Philadelphia, Black Hawk's company was taken to the Fairmount waterworks—a technological achievement that supplied the city with a constant and bountiful supply of clean water. In Albany, they were shown the arsenal—a reminder of the superior weaponry that defeated them.[99] One poet, addressing Black Hawk, noted that "There's much thou'st seen that must excite thy wonder."

> Our big canons with white and wide spread wings,
> That sweep the waters as birds sweep the sky;—
> Out steamboats with their iron lungs like things
> Of breathing life, that dash and hurray by?
> Or if thou scorn'st the wonders of the ocean,
> What think's thou of our railroad locomotion?[100]

Black Hawk and his fellow Sauk were both observers and observed. Spectators jostled for position to get a view of the "wild warriors."[101] Newspapers reported the captives' reaction to the "novel sight of a balloon ascension. . . . It is said that they exhibited greater astonishment on this occasion, than on any other during their route."[102] One reporter smugly asserted "It seems they are tired of the noise and bustle of our large cities, and long for the solitude of the wilderness. Thither let them go.—They have already seen enough to convince them that they have very little chance of *subduing* the United States, and consequently that any further encroachments on their part must result, as all previous ones have done, in their defeat and greater depression."[103] Durant's poem, composed for the occasion, reinforced the conviction that Native Americans were destined to be defeated by a superior race:

> *He* fought for *Independence* too— . . .
> But fought in vain—for 'tis decreed,
> His race must fall, and yours succeed. . . .
> He knows your strength—has felt your power—
> Then send him to his native bower.[104]

Not all commentators echoed this jingoistic rhetoric. David Claypoole Johnston, a popular visual satirist, published his illustration "The Grand National

Caravan Moving East" soon after Black Hawk's eastern journey. In Clay-poole's depiction, Black Hawk and his fellow captives ride in a caged wagon, towed along in a procession led by Andrew Jackson. Jackson acknowledges the cheering crowd while a balloon named "Rising Generation" carries an occupant, meant to be Durant, waving an American flag.[105]

While Americans deployed balloons as symbols of American enterprise and as demonstrations of sovereignty over conquered peoples, satirists and politicians made use of balloons to express both ambitions and failures. If audiences were well-prepared for Poe's hoax, they were equally ready to embrace the biting political satires that came thick and fast from the 1830s on when the advent of lithography enabled finely drawn mass-produced prints.[106] President Andrew Jackson and all subsequent politicians in the antebellum era were continually subjected to acerbic cartoons. Artists had always used scenes and symbols that resonated with viewers. In the 1830s and 1840s, some of the most salient images included air balloons. Balloons symbolized the folly of political ambitions, high-flown hopes, and abject fail-ures. Edward Williams Clay's lithograph "The Times" (1837) depicts the mis-ery inflicted on Americans by the Panic of 1837, an economic catastrophe triggered by Andrew Jackson's abolition of the United States Bank. Clay's political cartoon blames the Democrats, and specifically Jackson's policies.

FIGURE 2.4. David Claypoole Johnston, "The Grand National Caravan Moving East" (New York: Endicott and Swett, 1833). Library of Congress, Prints and Photographs Division.

In Clay's illustration, presidential policy results in misery and poverty for working-class Americans. Clay's street scene shows an inebriated family: the father holds out a bottle of gin and the mother, lying on the ground with her child, reaches for the bottle. A well-dressed woman and child beg from a prosperous banker with "bonds and mortgages" tucked under his arm, while the pawnbroker and liquor store are busy. Across the river are a debtors' prison and an almshouse. In the sky above this dismal scene, a punctured balloon labeled "Safety Fund," a reference to Martin Van Buren's insurance program for New York banks, plummets to the ground, its occupants tumbling from the basket.[107]

The political cartoonist H. Bucholzer commented on Whig aspirations to the presidency in 1844 with two cartoon prints, "Balloon ascension to the Presidential Chair" and "Bursting the Balloon." The first print shows the Whig candidate Henry Clay and his running mate Theodore Frelinghuysen ascending toward a presidential chair floating in the sky. Clay waves the American flag, and both men doff their hats at the Democrat James Polk and his running mate George Dallas standing below. Andrew Jackson vainly attempts to prop up Polk's deflated balloon, while Martin Van Buren, depicted as a fox, slinks away with the comment "If you had had the wit to put me in there, it would have gone up." In Bucholzer's companion print,

FIGURE 2.5.   Edward Williams Clay, "The Times" (New York: Printed and published by H. R. Robinson, 1837). Library of Congress, Prints and Photographs Division.

FIGURE 2.6. H. Bucholzer, "Balloon ascension to the Presidential Chair" (New York: Lith. and pub. by James Baillie, 1844). Library of Congress, Prints and Photographs Division.

"Bursting the Balloon," Clay and Frelinghuysen continue to rise while Polk and Dallas tumble out of their burst balloon. Perhaps inspired by these pro-Whig political cartoons, in August 1844 the Whig Party staged a balloon stunt.[108] Two balloons, one named Clay and the other named Polk, were released simultaneously. The Clay had been booby-trapped so that it would quickly deflate while the Polk soared away. To the dismay of the Polk supporters, the trap failed: Clay outstripped Polk and soared "majestically far upward."[109] As one reporter commented, "The next time our Whig friends undertake to get up an omen they should be careful not to trust their tricks to the chances of fate."[110]

"The Way They Go to California"

While politicians adopted balloons to float their rising ambitions, and satirists used balloons to signify crushed political hopes, the inventor Rufus Porter envisioned practical balloon travel to achieve continental settlement. The play *Ladies Rights!* presumed Americans would have to wait a century for rapid transcontinental flights. But Porter designed, and began to build, an airship to transport eager prospectors to the California goldfields in 1849. No American believed in the promise of technology more than Porter. An itinerant mural painter, his passion for invention led to dozens of patents for items of utility, including a fire alarm, life preserver, cheese press, fog whistle, punching machine, automatic grain-weighing machine, and a pre-fabricated movable house.[111] He developed a revolving rifle which he sold to Colt in 1844. His later patents included a steam engine (1858) and an air pump (1863).

Porter wished to harness the brain power and enthusiasm of American workers and craftsmen. To aid the spread of information about new technologies and methods, he published a weekly newspaper, the *New York Mechanic* (1841–43), that reported the latest developments.[112] Four years later, Porter began the *Scientific American* (1845), a publication that brought technological developments to a national audience. Porter's goal in publishing the monthly journal was to publicize "more extensive intelligence in Arts and Trades in general, but more particularly in the several new, curious and useful arts, which have but recently been discovered and introduced."[113] The title of the journal expressed inclusiveness. Porter's imagined readership was not simply inventors, mechanics, and tinkerers. Americans were a scientific people. Subscription numbers bore out Porter's assumption.

In the second issue, Porter supplied readers with a detailed illustration of his "Travelling Balloon." Porter assured his readers that it was a safe way to travel, safer than sailing ships, steamboats, or railroad locomotives. "The steam engine by which the balloon is propelled, will be very small, and the boiler being constructed of small copper tubes, there can be no possibility of damage by explosion." And just in case, each passenger would be issued a parachute.[114] Four years later, Porter had refined and improved his ship. Now he began promoting it in earnest. His pamphlet *Aerial Navigation* (1849) assured potential investors and customers that a trip from New York to California in Porter's "Aerial Locomotive" would be safe and enjoyable. Moreover, Americans could take pride in surpassing the "proud nations of Europe [who would be] staring and wondering at the soaring enterprise of the independent citizens of the United States."[115]

Scientific American regularly reported advances in aerial travel along with numerous inventions for more efficient gold extractions. In March 1849, for instance, the magazine had a section titled "California Inventions." The lead article illustrated a gold dredging and washing machine.[116] Gold fever kept public interest in Porter's airship alive. In 1851 Porter and his partner, a Mr. Robjohn, began building the full-size machine in Hoboken, New Jersey. The popular family magazine, *Gleason's Pictorial Drawing—Room Companion*, followed the ship's progress and provided readers with a glimpse of the Aerial Locomotive's passenger car under construction.[117] But Porter was

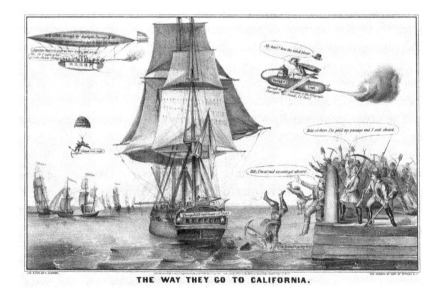

THE WAY THEY GO TO CALIFORNIA.

FIGURE 2.7.   "The Way They Go to California" (Spruce, NY: N. Currier, c. 1849). Library of Congress, Prints and Photographs Division.

constantly short of funds to complete his ship. A request from Congress for five thousand dollars was denied, despite the fact that Porter emphasized the national utility of his machine—not just for conveying gold prospectors to California, but as a speedy method of transporting the US mail.[118] Porter's ambition to serve the nation, though thwarted by Congress, caught the public's imagination. A satirical print, "The Great Pictorial Romance of the Age, or Steam Ship Commodores & United States Mail Contractors," illustrated an airship that would transport the US mail.[119] Another print, "The Way They Go to California," shows Porter's Aerial Locomotive flying overhead while gold prospectors, who literally missed the boat, tumble into the harbor, and a man on another transportation alternative, the Rocket Line, flies over the miners' heads.[120] Porter was still refining the design of his airship in the 1860s, but the Aerial Locomotive was never completed.

Weapons of War

Just at the moment Porter abandoned his plan for transcontinental flight, war finally brought balloons into practical use. If balloons could not fulfill the hopes of Americans for rapid transportation and economic development, they might, so promoters argued, help the Union defeat the Confederacy. For those already interested in the possibilities of aerial technology, the war was an opportunity to put the technology to work. Joseph Henry, the first secretary of the Smithsonian Museum, promoted the interests of aerialists by bringing them to the attention of the Union army. By the summer of 1861, the aeronaut Thaddeus Lowe launched tethered balloons to observe Confederate encampments in Northern Virginia. One of Lowes's army balloons, fittingly, was named *Union*.[121]

The newly created army Air Balloon Corps operated seven balloons by the spring of 1862. The balloons were named to emphasize the Northern cause; the *Union*, *Constitution*, *United States*, and *Washington* all took to the air as visual reminders of the nation. Whenever *Harper's Weekly* magazine published an illustration of an Air Corps' balloon, the name was always visible.[122] Aerial observation proved effective: the Union army's view of enemy encampments and troop movements compelled the Confederates to employ new strategies to hide their activities from enemy eyes. They built false campfires and fake artillery batteries in the hopes that the Union army would misperceive what Confederate troops were up to.[123] The Balloon Corps' success in uncovering Confederate activities may have inspired one image used on illustrated Civil War envelopes sold in the North: it showed men tumbling from the basket of a deflated balloon labeled "Secession."[124]

The importance of balloons as a weapon to win the war was not lost on the general public. Northern newspapers kept their readers informed about every aspect of the conflict, including the technologies employed to defeat the rebels. Reports of engineering feats—new bridges to carry transport and troops, and improvements in weaponry filled the pages of Northern newspapers and magazines. The *Harper's Weekly* illustration "Balloon View of Washington D.C." showed buildings, streets, and details of the landscape in sharp detail, thus confirming for readers the benefits of observing the enemy from the air. In Currier and Ives's evocative print of the Battle of Fair Oaks in May 1862, Lowes's observation balloon hovers above the clashing troops in the upper left corner.[125] The following month, the *New York Times* informed readers that General McClellan ascended in Lowe's balloon (the *Union*) to observe Confederate positions in northern Virginia.[126] Six months later, the same paper covered most of the front page with a detailed map labeled "Balloon view of the Approach to Richmond and Rebel Defenses."[127] Readers were treated to a view of the Confederate capital that was only possible from overhead. The *Times* map showed readers how accurate balloon observation was; the article accompanying the illustration gave details of people moving around the city streets (and watching the balloon).[128] Visual evidence reinforced the battlefield reports. The numerous illustrations in *Harper's Weekly* gave readers a battlefield-side view. In some instances, the information relayed to the public was too accurate: Secretary of War Edwin M. Stanton ordered *Harper's Weekly* suspended after the magazine published two bird's-eye maps showing the position and identity of Union infantry and McClellan's headquarters. Stanton declared the magazine was "guilty of giving aid and comfort to the enemy."[129] The air balloon's contribution to the war did not go unnoticed by cartoonists. Many civilians expressed frustration with what they perceived as McClellan's hesitation to attack Richmond in 1862. *Frank Leslie's Illustrated Newspaper* offered up a suggestion to speed the Union army along: "Flying Artillery—a Hint to General McClellan how to 'advance on Richmond'" shows an artillery squad flying toward Richmond in a balloon.[130]

Though balloons were as yet incapable of conveying troops, they proved their strategic value when Thaddeus Lowe convinced the War Department that it was possible to relay observations from the tethered balloons to commanders in the field by telegraph. Lowe had earlier made an ascension with a telegraph cable from the grounds of the Smithsonian in June 1861. To drive

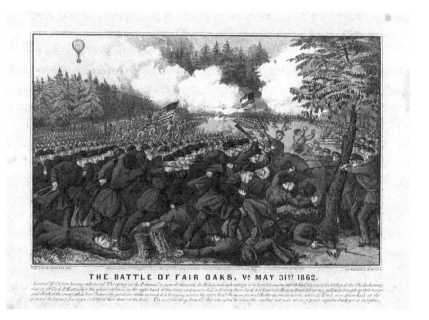

FIGURE 2.8. Currier and Ives, "Battle of Fair Oaks, Va., May 31st, 1862." Smithsonian National Air and Space Museum.

home his point to President Lincoln, Lowe sent a telegraph message from his balloon to the White House.[131] Soon the newspapers reported the effectiveness of combining these two new technologies. Journalists expressed excitement at putting them to work. In one *New York Times* article headlined "Morse Electric Telegraph: Its Utility to General McClellan," Parker Spring, a Western Union telegrapher, gave readers a breathless account of the risk aerialists took by their proximity to rebels:

> As the fight progresses, hasty observations were made by the Professor and given to me verbally, all of which I instantly forwarded to Gen. MCCLELLAN and Division Commanders through the agency of the obedient field instrument, which stood by our side in the bottom of the car. Occasionally a masked rebel battery would open upon our brave fellows. In such cases the occupants of the balloon would inform our artillerists of its position, and the next shot or two would, in every case, silence the masked and annoying customers. For hours, and until quite dark, we remained in the air, the telegraph keeping up constant communication with some point. From the balloon to Fortress Monroe, a distance of over 100 miles, this wire worked beautifully.[132]

Spring provided details about the streets of Richmond (mostly deserted), the army removing dead and wounded, and Confederate troops marching out of the city. He emphasized the value of these new technologies to the Union effort by evoking the American revolutionaries' cause: "PATRICK HENRY, I believe it was, who said, 'We must make use of those means which God and Nature has placed in our hands.' We are doing it."[133]

Despite ample proof that air balloons were an effective weapon of war, by the spring of 1863 the Union had abandoned balloon observations and disbanded the Balloon Corps. There were several causes for the failure to continue balloon surveillance, including infighting among aerialists vying for government money. And Lowe was never given a military appointment. He was merely a civilian adviser—an uncertain, and as it turned out, tenuous position. The military command failed to use the information obtained from aerial observations effectively. With no clear mission, and no direct chain of decision making, aerial observations failed to justify further expenditures.[134] Nevertheless, air balloons remained a symbol of successful Union efforts to win the war. Stanley and Conant's *Polemorama, or Gigantic Illustrations of the War* exhibited in Boston, displayed key scenes from the conflict, including "Balloon Reconnaissance at Bailey's Crossroads."[135] A popular print of General Grant, published in 1864, showed the war hero surrounded by images of crucial moments in the conflict and by an observation balloon.[136] The aeronaut Charles A. Stearns drew an audience to his Fourth of July ascension the same year by naming his balloon "General Grant" and distributing H. F. Durant's "National Ode" to the crowd. Durant's poem celebrated the balloon as a symbol of the Union's triumph:

> Thou art our emblem! Our purpose has been,
> And is, to be still a Nation of Men;
> To arise in our majesty godlike and free,
> As thy form is uplifted o'er land and o'er sea;
> To bear the glad tidings of union and strength
> All over the land, through its breadth and its length:
> Union in justice and hatred of wrong,
> Strength over tyranny, however strong![137]

Ironically, the air balloon's failure to fulfill any of the hoped-for improvements—whether in transportation or in warfare—did nothing to dampen American enthusiasm for the technology. Balloons functioned as a touchstone for American ambitions. Through balloons, Americans articulated national aspirations and national pride. A medal cast for a Providence,

Rhode Island, high school in 1829 illustrates just one of the ways Americans put balloons to symbolic use: illustrations on the medal included "a beehive, emblematic of industry, a balloon, of progress upward, a harp, very delicately wrought, together with globes, books, writing apparatus, &c."[138] Balloons could not ship grain to market, transport miners to California, or deliver the US mail, but balloons could carry the United States to its destiny. Without even leaving the ground, balloons served Americans well.

CHAPTER 3

Engines of Change

Machines Drive American Industry

A national project of improvement was under way in the early republic. Assisting in this endeavor were new societies formed to encourage American invention. One of the earliest of these was the Delaware Society for the Encouragement and Promotion of the Manufactures of the United States of America. Begun in 1788, the society took American independence as its creed. The formal statement of the society's goals echoed the language of the Declaration of Independence: "The author of wisdom has endowed them [Americans] with virtue, patriotism, and a just sense of the rights and dignity of man." It was, therefore, the duty of the "sons of America" to promote the arts and sciences and to increase manufacturing. At the same time, the society aimed to "discourage the importation of foreign articles, and always give a preference to domestic manufactures, when there is a reasonable proportion between their prices and goodness." As a measure of commitment to the goal of material independence, every year on January 1 the members were to appear "in a full and compleat suit of American manufactures."[1] Organizations like the Delaware Society proliferated in the early nineteenth century. They were joined in the promotion of American invention by mechanics' institutes and lyceums that offered lectures and courses of instruction to a broad spectrum of American society. Itinerant lecturers traveled in the far South and the West, reaching even more Americans in newly settled areas. This background of information and experiences built an audience for engines of change. The steam engine arrived in America before the nation declared political independence. Although private investors took time to build enthusiasm and promotion for the technology, once they did, steam was embraced as the essential motive power for private enterprise and a necessary aid to the national project of development. Steam-driven machinery benefited mechanics, merchants, and farmers. Goods and people traveled American waterways and railways with ease and speed. But the technology that most enthralled the nation in the early years of the republic was perpetual motion. A power source that did not require to be fed, pushed, or wound up had long been the dream of mechanics and inventors. Perpetual motion machines captured the public's imagination as the perfect engine for American ambitions.

The Perfect Engine

In April 1812, newspaper readers along the East Coast took note of an unusual announcement: Mr. Charles Redheffer claimed to have "discovered and for some time had in operation a self-moving machine or Perpetual Motion."[2] This was not the first time, nor would it be the last, that Americans were teased with the possibility of a machine that could run by itself. Redheffer's machine, exhibited at Chestnut Hill, just outside Philadelphia, was similar to others before it. But Redheffer was in the right place at the right time. Whereas other perpetual motion inventions had garnered a small share of public attention, Redheffer's device appeared at a moment when Americans were especially eager to develop agricultural production and manufacturing with improved machinery. The nation strove to fill the gap left by the withdrawal of British goods, first during the Revolution, then by the Embargo and the War of 1812. American ingenuity and enterprise were applauded and supported by various societies and organizations devoted to internal improvement. The steam-powered inventions of Oliver Evans, John Fitch, and Robert Fulton made possible more efficient mills, increased manufacturing, and better techniques for harnessing natural resources. All required energy—either animal, wind, or water. What if the energy, too, could be made to work harder, more efficiently, and continuously?

Perpetual motion was not a new idea. From the ancient world to the early modern era, individuals claimed to have discovered its secret. Devices generating an everlasting source of power had always had their champions. The machines were promoted as a benefit to mankind, as proof of national superiority, or most often, as surefire commercial investments.[3]

It is unsurprising that American inventors from Boston to Charleston strove to be the first to achieve perpetual motion. Almost every year between 1795 and 1816, someone announced that they had discovered the secret. A few machines were quickly exposed as fraudulent, a few reports were self-congratulatory puffs, while others stretched credulity. In February 1795, Americans learned of a perpetual motion machine exhibited in Providence, Rhode Island. Thanks to the skepticism of some of the audience, the truth was revealed: "on a *high* shelf, a boy was discovered (perdue) *turning an Iron Crank.*"[4] In Baltimore, John Stewart announced his invention with the following statement: "With supreme satisfaction and assurance I proclaim to the inhabitants of this new world, the infallible discovery of a Perpetual Motion."[5] In 1800 two prisoners in the Goshen, New York, jail allegedly built a perpetual motion machine "without any other instrument than a penknife."[6]

Some inventors emphasized their contribution to national development and productivity. David Moore, the inspector of flour for the Port of Baltimore, exclaimed to readers of the Philadelphia *Independent Gazetteer*, "Happy Columbia! My native country that possesses every soil and clime, and a genius that can command the stones of thy fields, and the trees of thy forests, to perform all the acts of tillage, and smooth water navigation!" Moore claimed his machine would "plow, sow, and mow—impel all carriages, or impede them, at the will of the conductor." He planned to petition Congress for "the exclusive right of working all machines by wedge screw, lever or dead weight."[7] An emphasis on patriotism was a common thread in the letters sent to Thomas Jefferson. Many inventors sought assistance, advice, or money from the president. David Launy, for example, addressed Jefferson in flattering terms: "As the father of american Independence, as the protector and best Judge of arts and sciences, and as the chief Magistrate of our florishing Empire, I fulfill the duty of a Citizen, who has the happiness to live under Your Wise Government."[8] Some, like Matthew Wilson, hoped to secure Jefferson's attention and sympathy, as "a true Republican an old Soldier & volintier & a Sincier friend to his Cuntery . . . I Look Up to Your house as a Son to a Father for a Littl assistance & I will bring you the power of Pressuar Dead or Alive if alive it will be Useful to the World."[9] Some explained their device in detail. Chauncey Hall's perpetual motion machine was based on magnetic attraction. Others requested a meeting with Jefferson rather than risk putting their plan on paper. Ambrose Bayley, who claimed to be "in low circumstance," asked for money to travel to the capital so that he could reveal the secret of his discovery.[10] Others hoped that a grateful nation would reward their efforts. Joseph O'Neil expected that his contribution to national prosperity would garner him "a handsome Reward" from the government.[11] In 1810 Dayton Leonard asked President Madison to persuade Congress to award Leonard three hundred dollars to complete his machine.[12] Whether they were old soldiers or true patriots, perpetual motion inventors marketed themselves as devoted to the nation.

Most inventors are known only by one letter or a brief newspaper notice. A few pursued a variety of opportunities to advertise their discovery and to achieve both attention and funding. Lewis Du Pré of South Carolina placed the following announcement in a Charleston newspaper in November 1801: "The Subscriber takes the earliest opportunity of announcing to the world, that, (after an arduous pursuit) he, this day, about half after twelve p.m. happily succeeded in the discovery of PERPETUAL MOTION. He has also the pleasure to add, that the principle is such as to be extensively useful in the different branches of Machinery."[13] To fund his project, Du Pré went straight to the top. In January 1802, he petitioned Congress for money to develop his invention. He explained that it would be impossible for him to reap the financial rewards of his discovery "without the aid of Legislative interference." Du Pré appealed to the representatives' "republican patriotism" to fund a project that would benefit the nation. He signed his letter "50th day of perfect motion, Jan. 1 1802."[14] Du Pré was confident that Congress would grant his request and award him funds and a patent, but he requested that Jefferson persuade Congress to extend the length of the patent beyond the fourteen years provided by the law. Du Pré again used patriotism in his appeal, emphasizing that his discovery had been made by "an *American*, for a *genuine republican* & that at a time when a *Jefferson* fill'd the presidential Chair."[15]

Confronted with Du Pré's "strange, disordered composition," Congressman Samuel Latham Mitchell of New York, a noted scientist and educator, assured his colleagues that "the object which the petitioner pretended to have attained, was contrary to the physical laws of matter. All experience and all philosophy was opposed to the notions of the kind contained in the paper before the House." Mitchell hoped Du Pré would receive no more attention from Congress. He dismissed the petition with the comment that "it was not worthy of the National Legislature to give serious attention to physical impossibilities." Nevertheless, after some debate, the legislators ordered that Du Pré's petition be forwarded to a select committee. This may have been a polite way of passing on the responsibility of rejecting Du Pré's scheme, since they appointed Mitchell to head the committee. Three weeks later, the record simply stated that Du Pré "[has] leave to withdraw his said petition."[16] In the meantime, Du Pré's activities caught the attention of the popular journal *Port Folio*, in which the editor referred to Du Pré as an "insane projector."[17] Insane or not, Du Pré's perpetual motion idea was mentioned by a national publication, and this interest may have served to increase American curiosity about perpetual motion. In 1804 *The Columbian Almanac* instructed readers in how to create perpetual motion for themselves using a lodestone in a bottle.[18] Two years later, at Boston's Columbian Museum, audiences were entertained by a demonstration of "a curious philosophical experiment, The Perpetual Motion."[19] Unlimited motive power was a goal that many still hoped to achieve.

Although Congress declined to support Du Pré, and Jefferson ignored most of the letters he received from inventors of perpetual motion, in 1812 Charles Redheffer succeeded in capturing a large share of the public's attention. In April that year, the Philadelphia *Democratic Press* announced his self-moving machine: "Mr. R has the most perfect conviction that the motion of the machine will never cease so long as the materials of which it is composed will last. He has exhibited it to many of his neighbors, all of whom express their astonishment at the perpetuity of its motion."[20] Redheffer charged one dollar to view his miraculous device. Curiosity was so intense that the merchant Thomas Cope believed Redheffer earned fifty to one hundred dollars a day.[21] By October, interest in his device was unabated. "P. T." in *Poulson's American Daily Advertiser* noted that there were many skeptics about the possibility of perpetual motion in general, and Redheffer's machine in particular. But, P. T. continued, "we would advise them, and all the curious, to see the inventor and the machine—and make themselves acquainted with the character of both." The search after perpetual motion, if unsuccessful, "has been productive of useful inventions and improvements in Mechanics, and there is no good reason to doubt, that it may be productive of many more." P. T. believed that "we ought to encourage and reward those who in the pursuit of it, discover any new and useful application of known powers in Mechanics."

William Duane, editor of the *Aurora General Advertiser*, was among those who took Redheffer's claim seriously. In an editorial explaining how the machine worked, Duane assured his readers that Redheffer's invention was worthy of their attention as a topic that pertained "to the honor and interests of the nation." Duane reminded his readers that Pennsylvanians had already "given to the world the demonstration of the identity of lightning and the electric fluid—that the *mariner's quadrant* was invented in Pennsylvania, that a native of Pennsylvania first powerfully applied the power of steam to navigation." The discovery of perpetual motion would be added to the state's accomplishments.[22] Duane claimed that Redheffer's machine was capable of powering "every species of labor-saving machinery." It would be a boon to manufacturing and agriculture: "all branches of carding, spinning, weaving, grinding of grain, or rolling of metals. . . . Neither heat nor cold affects or retards its operation; and the expense of construction will be within every man's power." Unlike the perpetual motion men who preceded him, Redheffer made headlines. He had what Du Pré and others did not—a champion. Thanks to Duane's enthusiasm, Redheffer's machine was news.

But there were doubters as well as supporters. The day after Duane published his panegyric, the *Aurora* reprinted a letter that was critical of Redheffer's device. The author, who identified himself only as "D. F.," claimed to have observed Redheffer's machine carefully and determined that it could not work. D. F. announced that he would build his own version to prove to the public's satisfaction that Redheffer's machine was a fraud. Another critic, who signed himself "Rittenhouse," led readers step by step through the laws of physics to show why Redheffer's machine could not generate perpetual motion.[23] A few days later, "Simplex," like "Rittenhouse," went point by point through the mechanics of motion. He, too, argued that Redheffer's machine was not what it claimed to be.[24] This public criticism only served to further acquaint readers with Redheffer's invention. More people talked about it and more people went to see it. Writing from Virginia, Thomas Jefferson told Robert Patterson, "We are full of it as far as this State, and I know not how much further."[25]

Patterson, director of the United States Mint in Philadelphia, sent Jefferson a broadside illustration of the machine.

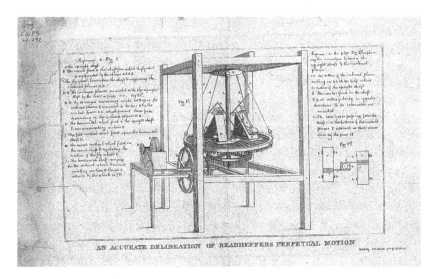

FIGURE 3.1. "An Accurate Delineation of Readheffers Perpetual Motion." Broadside, American Philosophical Society.

In his accompanying letter, Patterson complained to Jefferson, "It is mortifying to observe how many of our citizens, from whom better things might be expected, have become dupes to this imposture, and have even published their ignorance to the world." Patterson was mystified by the enthusiasm generated by Redheffer's machine. Even the Philadelphia City Council entertained the idea of putting it to work to raise water from the Schuylkill River (to replace the steam engine that powered the water pumps).[26]

While the debate continued over the truth or falsity of Redheffer's claim, the Pennsylvania legislature stepped into the fray. Redheffer had applied to the legislature for funds to develop his invention further.[27] In December 1812, the state House of Representatives appointed a committee to examine the machine. The legislature could have dismissed Redheffer's request out of hand, regarding it as unworthy of their attention and funding (as Congress had done with Du Pré). Instead, the legislators chose to act. If Redheffer's machine was indeed a successful perpetual motion device, the consequences were significant. The legislators reasoned that "not only great honor would be conferred upon the Commonwealth, but incalculable advantages would be derived from the invention by the people of the United States." Despite the naysayers, there was just a possibility that it might work. And if it did, the discovery of the ages would be credited to an American—and a Pennsylvanian. If, on the other hand, the machine was not what Redheffer claimed it was, "the public interest would be promoted by exposing its fallacy."[28] If the machine was a deception, the public, many of whom had already parted with a dollar to view it, had a right to know the truth.

The committee charged with investigating Redheffer's machine included Robert Patterson and the expert mechanic Oliver Evans. The men made several unsuccessful attempts to see it. On one occasion, they arrived at Redheffer's house to find the building locked and no sign of the owner. But they were able to peer at the machine through a window. Coleman Sellers, son of one of the men appointed by the legislature, saw enough of it, along with information from the illustration available in the city print shops, to determine that the machine could not possibly be self-powering. What Sellers noticed was that the wear marks on the gear teeth were wrong: the gear that was supposed to be driving the mechanism was, instead, itself being driven by another gear which must have been powered by a concealed source.[29] Sellers's father, Nathan, commissioned the mechanic Isaiah Lukens to make an exact model of Redheffer's machine. Lukens's ten-inch version of Redheffer's six-foot machine was displayed at Peale's museum. Even the model drew "some hundreds of citizens."[30] Duane confidently told his readers that Lukens's doubts about Redheffer's accomplishment compelled him to follow

Redheffer's design (based on the readily available illustration) and put it to the test. Lukens, Duane happily announced, "has had the justice to acknowledge the conviction of his first error." A bit smugly, he added, "and it remains now with the logicians and loggerheads—the mathematicians and meritricians—to open new batteries of folly and prejudice, and array all the artillery of paradox—to batter down the solid foundations of common sense."[31]

But Lukens's model was not a perpetual motion machine. Based on Sellers's conviction that a hidden source powered Redheffer's device, Lukens concealed his model's source of power in a false bottom.[32] It was not long before the truth was made known. Duane was livid, charging that Lukens "intended to weaken the public confidence in Mr. Readhefer's [sic] discovery and to *bilk the public* of a few dollars through the general aptitude of society to be *gulled* by the fraudulent, and to be *jealous and intolerant* of men of genius."[33] Robert Patterson revealed the secret of Lukens's model to Jefferson. The model "had the source of motion so artfully concealed, that it deceived (as the maker intended) Mr Duane, Mr Matlock, and many others, of the believers, as they are now termed, in Readhefers perpetual motion. Mr Lukins candidly confessed to me, the day after his machine was put in motion, that it was a deception, which he would discover after he had diverte[d] himself a little with the credulity of R's *believers*." Patterson closed his letter by asking Jefferson to excuse him for "saying so much on this silly subject. It being at present almost the sole topic of conversation amongst us must be my apology."[34]

As speculation and curiosity built regarding Redheffer's machine, another perpetual motion device went on exhibit in New York City. In early January, the *Columbian* announced that a perpetual motion machine could be viewed at the upper end of Broadway from 10 a.m. to 4 p.m. daily.[35] The curious paid one dollar each to visit this latest claimant to perpetual motion. But just as Redheffer's machine had its doubters, the New York device also attracted skeptics. Philadelphia had its share of expert mechanics and natural philosophers, but New York had Robert Fulton. At the request of several friends eager to know whether the machine was what it claimed to be, Fulton visited the machine in the company of the illustrious naval commander Stephen Decatur Jr. Decatur, recently lauded for his capture of the HMS *Macedonian*, was now in command of a naval squadron protecting New York harbor.[36] Within a few minutes, Fulton recognized that the mechanism was powered by a crank. Fulton then proceeded to dismantle part of the machine, uncovering a thin catgut belt that ran through the wall and up to the attic. There, he discovered a man turning the crank that powered the machine in the room below. The newspapers related this dramatic exposure: the *New York*

Evening Post screamed "Perpetual motion—All's over!—the secret's out!"[37] The episode had all the elements of a great story: the illustrious inventor and the naval hero as the champions of truth, and the charlatan who bilked the public of money and challenged Redheffer as claimant to the discovery of perpetual motion. Reports embellished the scene with details: Fulton was described as "the ingenious inventor of steamboats and torpedoes."[38] The discovery in the attic included "a poor old wretch with an immense beard, and all the appearance of having suffered a long imprisonment; who, when they broke in upon him, was unconscious of what had happened below, and who, while he was seated on a stool, gnawing a crust, was with one hand turning a crank."[39] The "enraged spectators," like the disappointed crowd at the Philadelphia balloon launch, demolished the fraudulent machine.[40]

The New York deception occupied the public's mind while they waited for the Pennsylvania committee to examine Redheffer's device and announce their verdict. In the meantime, William Duane, still Redheffer's champion, asked the public to suspend their judgment.[41] But Redheffer gave the committee one excuse after another for canceling the arranged meeting. By January 25 even Duane had publicly expressed his doubts. He told his *Aurora* readers that even though he had witnessed "a considerable number of respectable millwrights, and others conversant in machinery" disassemble the machine and found nothing to indicate there was any concealment, Redheffer's refusal to exhibit the machine "though it does not *prove* any deception in the machine, is nevertheless too mysterious and unreasonable to assure confidence, and unless explained in an open and unequivocal manner, must shake the judgement of those who felt the greatest gratification in the belief of its reality."[42] Duane was backpedaling.

The *Port Folio*, a magazine with a national readership, devoted several pages of the February issue to Redheffer's machine. The pseudonymous "Rittenhouse" contributed a long, detailed explanation of why the mechanism could not be perpetual. The editor, less dismissive of Redheffer than he had been of Du Pré, noted that perpetual motion was "every where heard of, and discussed, and assailed, or reprobated." He credited the "ardent, inquisitive spirit of our countrymen," for the fact that "the great question of perpetual motion has occupied and absorbed the public attention." Even science-minded women, many of whom benefited from science lessons at young ladies academies, or public lectures, applied their knowledge of "the principles of attraction, the effects of gravity, the power of plane inclinations, and the vain endeavours of untwisting chains." Despite the fact that the United States was at war, even politics was overshadowed by perpetual motion: "The war with England, and the invasion of Canada, possess only a subordinate interest in comparison with Mr. Readhefer [*sic*], and the Germantown discovery." This may have been a slight exaggeration, since the illustrations for the February *Port Folio* included both a drawing of Redheffer's machine and the battle between the USS *Wasp* and the HMS *Frolic* (the *Wasp* had successfully captured the British warship the previous year). The conflict with Britain was never far from the public mind, even if science provided a temporary diversion.

On January 28, the committee reported to the legislature that they believed Redheffer's machine was a deception. They gave their reasons: "The refusal of the said Redheffer to permit them to examine the same; and, 'that from the conduct of Charles Redheffer, as well as from numerous vain attempts to construct self-moving machines on the ostensible principles of his, it is, their decided opinion, that Charles Redheffer's machine of pretended perpetual motion is a deception, and himself an imposter."[43] The announcement was probably not a surprise. Redheffer's behavior had been highly suspicious, Robert Patterson was well-versed in theoretical and practical mechanics, and Lukens's model and the fraud exposed in New York demonstrated methods by which Redheffer could have brought off a deception. All these factors sowed doubts in the minds of the committee. Opinion shifted quickly from support to derision. In Philadelphia, a comic paper, the *Tickler*, revealed that the secret of Redheffer's machine was a squirrel on a treadmill.[44] Some enterprising individuals took advantage of the fascination with perpetual motion to sell mills, kilns, medicine, and even boxing bouts. Nathaniel Brown's gristmill, "built on the principle of Perpetual Motion," was for sale to satisfy his debts in Washington, DC.[45] Thomas Power advertised a "Perpetual Lime Kiln" for sale in Hudson, New York.[46] Even the purveyors of a headache remedy managed to insert a reference to the current interest in Redheffer's machine; the "progress of public opinion" of the snuff sold by Thomas and Whipple in Salem, Massachusetts, was said to be more positive than the belief in perpetual motion.[47] And the "Perpetual Giant Boxer" charged Philadelphia gentlemen five dollars to get in the ring with him.[48]

But that was not the last the public heard from Redheffer. Two and half years later, he again drew national attention. In October 1815, a New York paper announced that Redheffer would soon submit his perpetual motion machine for public examination.[49] Critical and sarcastic comments immediately followed. The *Salem Gazette* thanked Redheffer for providing newspaper copy in lieu of anything truly newsworthy to report: "To help us out of the scrape, Mr. Redheffer is about setting his perpetual motion to work again. . . . Certainly a newspaper would not be worth reading, were not these small trifles sent to us as an offset for European battles, and such like matter as Bonaparte was wont to make for our politicians."[50] But "Viator," in the Philadelphia *Democratic Press*, asked the public to give Redheffer an opportunity to demonstrate his new machine.[51] Even William Duane offered Redheffer a second chance: "It is hoped that the asperity towards the machinist, which cast a shade on the liberality of society, will at least be suspended until there has been a fair and dispassionate investigation."[52] Unlike William Duane, the Pennsylvania legislature was not willing to give Redheffer a second chance. But the allure of unlimited energy compelled a number of Philadelphians to agree to Redheffer's request that they examine the new machine. Robert Patterson and Nathan Sellers were again on the committee. Leading men in the city and the state, including the mayor Robert Wharton; the chief justice of the Supreme Court of Pennsylvania William Tilghman; General Thomas Cadwallader; Congressman Joseph Hopkinson; and the lawyer and former congressman Charles J. Ingersoll, agreed to serve.[53]

Redheffer disassembled his machine and brought it to a room in the west wing of the State House where he planned to reassemble it under the watchful eyes of the committee, and to demonstrate how it worked. On Saturday, July 27, 1816, the committee, joined by half a dozen more gentlemen invited by Redheffer, watched as he began to assemble the machine. By two o'clock in the afternoon, the machine was still not complete. Redheffer took Mayor Wharton aside and explained that the delay was due to damage the machine had suffered on its journey from Chestnut Hill to the State House. He asked for a postponement. The committee gave Redheffer a week to make repairs. The following Saturday, the committee again assembled in the State House. This time, Redheffer requested that only a few of the men, preferably "those persons of known probity and experience in mechanics," be present when he demonstrated how the machine worked. Redheffer explained that he had not yet applied for a patent, and he wished to protect his financial interest in the machine.[54] The committee assured Redheffer that they had no wish to learn the machine's secrets, but merely see it assembled and put in motion. But Redheffer refused to do so. The meeting ended, and Redheffer took his machine back to Chestnut Hill. But the committee did not disperse. Frustrated, and perhaps embarrassed by Redheffer's refusal to demonstrate his machine to the august company after so much buildup in the press, the committee composed a long letter detailing their meetings and Redheffer's inexplicable behavior. The letter was signed by all the members and sent to the newspapers. It stated that "the undersigned, therefore, withdraw from any attendance on Mr. Redheffer, with strong sentiments of disapprobation of his conduct. Of the feasibility of his project, no evidence has been given by him; and what inference should be drawn of his own belief in it, may be made by the public from his whole conduct on this occasion."[55] Tilghman and the others meant to distance themselves from an increasingly likely fraud. Redheffer's longtime champion, William Duane, remained silent. All of this was negative publicity for the state of Pennsylvania. One New York newspaper commented that Pennsylvania had "made herself conspicuous for her Whiskey Insurrections, Perpetual Motions, Ann Carson conspiracies, and Banking Institutions."[56]

Always in pursuit of a good story, the press continued to follow news of other perpetual motion machines. James Kirkpatric of Harrodsburg, Kentucky, announced that his machine could power two sawmills and six gristmills at once. The machine was so easy to operate that "a boy of 15 years old can put the whole in operation at pleasure."[57] William Foster of Broome County, New York, exhibited his machine in Geneva, Auburn, and Owego, where it was examined and dissected by "several of the most respectable citizens of this village [Owego]," who could not find any deception.[58] Perhaps emulating Redheffer's attempts to interest the state in funding his invention, Foster planned to take his machine to Albany, "during the present session of the legislature, to exhibit for the gratification of the curious and the learned."[59] Like Redheffer, Foster had his champions. The *Plough Boy* proclaimed, "It is believed that this invention is eminently calculated to benefit the world, and must immortalize the name of its inventor."[60] Another New Yorker, Gilbert Grotecloss, drew on patriotism to stir interest in his perpetual motion machine. He likened perpetual motion to republican principles of equality. He humbly told his readers that he was "one of your plain, homemade republican brethren."[61] Such rhetorical efforts to tie invention to the good of the nation may explain why the press continued to take note of perpetual motion claims. Or perhaps, like the *Salem Gazette*, editors appreciated any news item that filled up a lull in political or martial events.

This desire for entertainment may explain why Redheffer continued to draw a share of the public's attention. Now more a curiosity than a possibility, his machine became one more exhibit in Charles Willson Peale's museum. In March 1817, Peale announced three new exhibits: Isaiah Lukens's model of Redheffer's invention, the hand of an Egyptian mummy, and an ivory tobacco stopper that had belonged to Oliver Cromwell.[62] A few blocks away on Front Street, customers of the dry goods merchant Conrad Sparhawk and Company viewed Redheffer's original machine, as well as "the one erected for the inspection of the corporation."[63] Redheffer did not need his machine because he was building a new one. In November 1819, he was once again exhibiting.[64] But there was no fanfare or controversy this time. The public's interest in perpetual motion had run down. The brief notice of Redheffer's death two years later merely noted that he was "celebrated for his attempts to invent a perpetual motion machine."[65] By the third decade of the nineteenth century, Redheffer was remembered not for his potential achievement, but for pulling a ruse. In 1827 a report of one Greensbury Baxter, who scammed the citizens of New York City out of one thousand dollars, was headlined "Redheffer Outdone!"[66]

The question asked about all the perpetual motion inventions through the ages has been, did their makers know that they did not work? For machines in the distant past, it is impossible to have a conclusive answer. But in the case of Charles Redheffer, the publicity surrounding his machine, and the scrutiny brought to bear on its construction, allow us to give a fairly conclusive answer. Redheffer's machine did not work, and he knew it.[67] Why then, did so many individuals announce that they had found the secret to perpetual motion? Why did the public pay attention to them, go to see them, and even support them, as Duane did with Redheffer? Public interest in a miraculous machine meant that people would pay to see it. The Pennsylvania legislature's attention to Redheffer's machine only added to the public's curiosity. And there was an outside chance that one of the machines might really work. But once a machine was shown to be an imposture, the public moved on to the next promising invention. In the early nineteenth century, that invention was the steam engine.

American Steam

In September 1819, the Reverend William Bentley, minister of Salem's Second Congregational Church, wrote in his diary, "Yesterday I rode to Nahant with Miss H., the intended Mrs. Kittredge, & Miss Dodge of N. Port. We visited the usual places & found a greater variety of Company. No steamboat or Sea Serpent."[68] This unlikely pairing of events made perfect sense to Bentley. Both the Gloucester Sea Serpent and steamboats were fascinating novelties to the ever-curious sixty-year-old.[69] Bentley kept an extensive diary from the 1780s until his death in December 1819. He routinely noted local, national, and international occurrences. He also recorded unusual events: Bentley marveled at the Pig of Knowledge, observed Blanchard's balloon ascension in Boston, attended electrical demonstrations, and dismissed Perkin's metallic tractors as nonsense. Bentley is a wonderful example of American curiosity about the natural and the manmade world. At the end of his life, the technology that captured his attention was the steam engine.

William Bentley's numerous diary entries on the steamboats he saw plying the waters between Salem and Boston reflect the fascination this new mode of transportation held for many Americans.

Steam captured the public's attention in a profound a way. Steam engines evoked an enthusiasm that was evident in the large crowds at demonstrations and boat launches, in the poetry and fiction about the new technology, and in goods for the home, including dinner plates and wallpaper. Much of this enthusiasm about steam was expressed in terms of ambitions for the nation—the plans, schemes, and designs for economic strength and productivity. Engines that drove machines, moved boats, and propelled locomotives were symbols of national achievement. Enthusiasm for these applications of steam technology was not limited to mechanics, engineers, and inventors. Expanding communication networks and new educational opportunities exposed a good portion of the population to the practical applications of the technology. Many Americans read Dionysius Lardner's popular texts on steam engines and attended his lectures throughout the United States.[70] Popular periodicals such as *Columbian Magazine* and *Port Folio* published essays on the steam engine. *Godey's Lady's Book* reviewed publications on steam technology and carried advertisements for steamboat travel.

The first steam engines were imported to colonial America from Britain. The Englishman James Watt marketed the first commercially available engine in 1776, but even before his engines were sold in Britain, these engines were demonstrated to interested audiences in the American colonies.[71] The technology of the steam-powered engine was fairly simple, but an understanding of the mechanics behind it, as well as a good amount of money, was required to build and maintain one. Despite these obstacles, steam captured the attention of colonials. Steam technology was a liberating innovation because steam engines could replace water-driven machines. Instead of machines going to a power source, the power source could be constructed anywhere. Not just curiosity, but the necessity for domestic manufacturing in the postrevolutionary era accelerated the search for technologies to enhance production. Tench Coxe, the most vocal advocate for American manufacturing, predicted in 1787 that steam-powered mills would soon appear in New England and elsewhere. He was not wrong.[72] But steam engines were expensive. Only a handful of Americans with private means could import a British engine and the mechanics who knew how to keep them running. Philip Schuyler purchased an engine for pumping water at his New Jersey copper mine in 1793. When Schuyler's partner, Nicholas Roosevelt, purchased an iron foundry across the river from Schuyler's mine, he hired two English engineers to build a steam engine for his manufacturing building.[73] Schuyler and Roosevelt were innovative entrepreneurs, but more than a decade passed before other businessmen adopted the new technology.

New York City was the first municipality to invest in steam. In 1774 the Common Council gave Christopher Colles 2,600 pounds to purchase a steam engine and to construct fourteen miles of pipes for the city. Unfortunately for Colles, and for the city's residents, the pumping system was completely neglected during the British occupation. After the war, the municipality left it to private water companies to supply the city's needs.[74]

More than a quarter of a century passed before another American city commissioned a steam-powered water supply: the engineer Benjamin Henry Latrobe's water system for Philadelphia opened in 1801. Though not the first steam-powered water system, it was the largest and most lasting. Water from the Schuylkill River was pumped to a distributing tower in Centre Square and from there through wooden pipes to hydrants throughout the city. Two steam engines, designed by Nicholas Roosevelt, supplied residents (with many interruptions) with fresh water. The Philadelphia Water Works proved conclusively that steam engines could replace water power.

Meanwhile, inventors had been at work for almost two decades developing a marine engine to take the place of wind power. In August 1787 John Fitch gave a successful demonstration of his steamboat, the *Perseverance*, to members of the Constitutional Convention in Philadelphia. This exhibition, before such an illustrious audience, was good advertising for Fitch's enterprise. The following summer he again demonstrated his steamboat, this time making a twenty-mile journey up the Delaware from Philadelphia to Burlington, New Jersey. According to one account of the event, men and women "assembled at all the prominent points along the river to see her pass." At Burlington, the wharf was crowded with people waiting for the *Perseverance* to arrive.[75] Fitch briefly opened a passenger service between Philadelphia and Burlington in the summer of 1790.[76] As exciting as the new technology was, Fitch's boat service was neither faster nor cheaper than stagecoach service along the river. Fitch and other inventors understood all too well that the popular embrace of steam travel came at a price—speed, dependability, and affordability were the keys to success. Fitch argued that steam engines would be "of infinite advantage to the United States."[77] Commercial profit was a yet-to-be-obtained goal, but Fitch's proof of the viability of steam-powered travel spurred the convention delegates—no doubt urged on by Tench Coxe of Pennsylvania—to include provision for government support (though not *financial* support) for developing technologies that would increase American production.[78] Like Alexander Hamilton, Coxe was worried about the nation's balance of trade. One remedy was to increase manufacturing. Coxe cogently argued that "an agricultural nation which exports its raw materials and imports its manufactures, never can be opulent, because every profitable advantage which can be derived from its productions is given into the hands of the manufacturing nation."[79]

Encouraging individuals to devise improved machines and processes was a necessary first step to national development. Article I, Section 8, of the Constitution ensured that the government would protect the rights of inventors by giving them exclusive rights to their inventions for a limited time. That this was a priority for the nation was demonstrated by the rapidity with which the first Congress passed the Patents Act in 1790.[80] Inventors could benefit financially from their ideas without the risk that someone else would instead profit from them. Fitch, and many other inventors, took advantage of the new patent law. There were so many patent applications in the first two years that in 1793 the law was revised to eliminate the required detailed examination of each application.[81] As one observer noted, "The whole world seems to be running patent mad."[82]

One man who hoped to profit from the patent law was Oliver Evans of Philadelphia. In 1790 Evans received one of only three patents granted in that first year for his automated milling machine. Evans's device was already in wide use in the Brandywine Valley region of Delaware and Pennsylvania. The patent gave Evans the protection he needed to profit from his innovative design. It also gave him the confidence to apply his ideas on engine power to steam technology. When the Philadelphia Board of Health decided to dredge the Delaware River in 1805, Evans easily won the contract. The machine he built, the *Oraktur Amphililos*, was the first steam-driven land vehicle. The *Oraktur* was a boat, but a boat with wheels. To demonstrate his achievement, Evans advertised in the Philadelphia newspapers that spectators could watch the *Oraktur* circling around the Centre Square pumphouse before he launched the dredger into the Schuylkill, from where he sailed it to the Delaware.[83] The neoclassical building at Centre Square housed a steam engine that pushed Schuylkill water throughout the city. Thus, Philadelphians were already familiar with the power and reliability of steam engines. Evans's choice of Centre Square for his public demonstration capitalized on this achievement. Evans went one step further by showing his curious spectators (who paid twenty-five cents) that steam could drive land transportation; steam power could replace horse power.[84] Evans predicted "the time will come when people will travel in stages moved by steam engines from one city to another almost as fast as birds can fly, fifteen or twenty miles an hour."[85] It would be several decades before steam-powered locomotives fulfilled Evans's forecast. In the meantime, steam engines proliferated, driving machinery for mills, boats, and industry. Evans's Mars Works (named after the god of war) manufactured many of them. By the 1810s, his Pittsburgh manufactory was producing engines for Mississippi riverboats. In 1812 Evans christened his newest engine the *Columbian*, intentionally giving a patriotic name to a machine that would be of service to the nation. Despite Evans's commercial success from his engine factories in Philadelphia and Pittsburgh, he felt that the patent laws had failed to protect his intellectual property.[86] Evans made repeated unsuccessful attempts to convince Congress to extend the time limit beyond the fourteen years allowed by the law. Without such protection, competitors quickly brought out their own Evans-designed machines.[87] Evans argued that the time limit injured him and all other American inventors. But it also injured the country; he told Congress that inventors wished "to serve our country in the way the God of nature has qualified us best to serve it."[88] Those who denied inventors their right of ownership through patent protection were, Evans claimed, the "enemies to the prosperity, wealth and power of the nation."[89]

Robert Fulton benefited from the work of both Fitch and Evans. In 1807 Fulton's *North River Steamboat of Clermont* traveled up the Hudson River from New York City to Albany. Almost no one had seen a steam-driven boat before: Fitch's Philadelphia boat service twenty years earlier had only lasted one season. Evans's *Oraktur* made a single voyage from the Schuylkill to the Delaware. The *Clermont*, on the other hand, drew an audience to the Hudson River for hundreds of miles. People lined the riverbank to gape at the "monster moving on the waters, defying the winds and tide, and breathing flames and smoke."[90] Fulton's friend Cadwallader Colden reported that crews from passing sailboats "prostrated themselves, and besought Providence to protect them from the approaches of the horrible monster, which was marching on the tides and lighting its path by the fires which it vomited."[91]

The *Clermont* made the trip from New York to Albany in thirty-two hours. By sailboat the journey took four days. The first successful commercial steamboat service was under way. Fulton told the press "The success of my experiment gives me great hope that such boats may be rendered of much importance to my country."[92] Patriotism was profitable. But Fulton realized, perhaps aware of the frustrations Oliver Evans had suffered, that there was little chance of a financial windfall from simply holding the patent on a steamboat design. The big money was in a monopoly of the transportation route. Confident in the success of steam engine technology, in 1803 Fulton and his partner, Robert R. Livingston, were granted exclusive rights (for a maximum of thirty years) to passenger transportation by steamboat on the Hudson River from New York City to Albany.[93] Four years before the *Clermont* embarked on her maiden voyage, Fulton was already poised to become a wealthy man. He next embarked on a successful steamboat voyage from Pittsburgh to New Orleans, down the Ohio and Mississippi Rivers. Though he could not hold a monopoly on Mississippi River traffic, his steamboats were proven to be up to the challenge of lengthy, and speedy, inland river transportation.

The Popular Response to Steam Technology

Americans had two principal responses to steam technology: first, celebration of the way steam engines enhanced daily life, from the pumps that supplied water for cities to industrial machinery and steam-driven transportation, and second, the expression of fantasies (or fears) of the future—what technological developments might lay ahead, for good or for ill. Praise and enthusiasm for steam power abounded in American newspapers. As steamboats and locomotives came into use, the wits and poets took the new technology as their subject. Robert Fulton was the first celebrated technology hero in the United States. The same summer that Fulton began his Hudson River steamboat service, the British warship *Leopard* attacked the USS *Chesapeake*. Fulton responded to this threat to the nation's security. His strategy was not to build a navy capable of meeting the British on an equal martial footing, but rather to destroy British ships that came within American waters. His system of torpedoes could do just that. Fulton saw this scheme as a patriotic enterprise: money was better spent on internal improvements, especially canals, and the development of manufacturing rather than on an expensive navy. A day's trial of his system proved that torpedoes worked, but not sufficiently well for the navy to development the technology.[94] Nevertheless, Fulton believed that such "useful improvements" were "the nation's glory."[95]

Fulton was celebrated as the defender of the nation. With literary saber rattling, one poem declared,

> Here's freedom and peace with our Gun boats in harbor,
> Defying those foes who would feast on our labour,
> In distant regions oppression is known,
> Men sigh for their liberty and seek a new home,
> In our land of improvements Artists are rising
> On board of the Steamboat they, pleasantly Steering, . . .
>
> With skill—raise—raise—Your Voices in defiance
> Of the haughty foe who dare invade our land,

While Fulton's Torpedoes and Steamboat
range our strand . . .

Fulton is a Genius of this our Nation,
Protect his designs in the Womb of creation, . . .

Then why should the sons of Columbia dispute,
Whilst in wisdom and science they gain their
repute.[96]

Another poem continued the martial theme:

Sons of Cincinnati as history do tell,
Brave warriors in the field—domestics by spell,
In the times of battle with their shield and their blades
Valiantly they fought and in peace mind their glades
Sons of Columbia as our fathers of old,
Mount—mount your pegassuses and your wings unfold,
As the Steam Boat in wonder out strips all in chase,
In praise of our genius, exult in the race.[97]

Steamboats were an example of the United States' martial power. But the poets also celebrated the steamboat (and Fulton) as examples of native genius fostered by liberty. Technology and innovation protected the country; as Cadwallader Colden asserted, "No longer will the citizens of one nation . . . be exposed to the insults of lawless power of another."[98]

Peacetime use of steam technology far outstripped warfare applications. By the 1820s passenger travel by steamboat had become commonplace, and a lighter strain of steamy verse graced the nation's newspapers. "The Steamboat" extolled the pleasures of the new transportation:

A fig for all your clumsy craft,
Your pleasure-boats and packets,
The steam-boat lands you soon and safe,
At Manfield's, Troot's or Brackett's. . . .

We dance, and drink, and slug and laugh,
Nor think that we are near land,
But so it is—too soon by half,
We get to Brackett's Island.

But, here, or there, it matters not,
For everything is handy;

My jolly boys—I'll tell you what—
The steam-boat is the DANDY![99]

Steamboats were celebrated in music as well as poetry. The "Alida Waltz" was published in 1847, the year that steamboat began service on the Hudson between New York City and Albany. Steamboats quickly became associated with pleasure and recreation; sheet music, such as the "Alida Waltz," was often purchased as a souvenir of an excursion.[100] Steamboats not only inspired music, they could produce it too. Thanks to J. C. Stoddard's patent, an "Apparatus for Producing Music by Steam or Compressed Air," steamboats became a musical instrument of sorts: by the 1850s, the calliope was tooting tunes up and down American rivers.[101]

Steamboats quickly became part of American life. Even the president of the United States traveled by steamboat.[102] Children played with toy steamboats and read stories about steamboat journeys.[103] Consumer goods were embellished with steamboat imagery: inexpensive bandboxes, tableware, and even wallpaper bore images of this modern mode of transportation. One dinner plate displayed the steamboat *Marshall*, named in honor of the United States chief justice who struck down Fulton's monopoly of the Hudson River (though praised as an innovator, ironically Fulton was accused of obstructing economic enterprise). Panoramic scenes on wallpaper covered reception rooms or halls in the homes of wealthy Americans. *Vues d'Amerique du nord*, which graced the Stoner house in Thurmont, Maryland, was a travelogue of scenic natural views. One section of the panorama displayed a steamboat at the foot of Niagara Falls.[104]

Transportation, manufacturing, and water systems all benefited from steam. So did entrepreneurs: one grain mill in Virginia, the Norfolk Steam Mill, advertised the technology in its name.[105] In Savannah, a selling point of William Bell and J. S. D. Montmollin's cotton gin was that it could be powered by steam.[106] As the demand for steam power increased, mechanics advertised their services: Taylor and Roebuck, millwrights and engine makers in Philadelphia, informed the public that they made steam engines "in the most modern approved manner."[107] Inevitably, purchasers of steam engines needed them repaired, and want ads appeared for those who knew how to fix them.[108] Steam was used to advertise fabric, hotels, stagecoach services, and real estate. Dry goods merchants sold "Steam-loomed shirtings."[109] Drawing on the public's fascination with another form of technology, one enterprising owner named his steamboat *Balloon*.[110] In 1810 Ann Wilkins informed prospective purchasers that the ten lots she had for sale in New York City were "the best situation in the city" because of their proximity to Fulton and Livingston's steamboat wharf.[111] The Steam Boat Hotel in Jersey City was located near the ferry where, the proprietors assured their guests, "large, convenient, and perfectly safe Steam Boats" docked.[112] The "Steam Boat Stages" between Baltimore and Philadelphia were simply stagecoaches, not steamboats. But advertising by association paid off—passengers could easily connect from one form of transportation to the other in order to complete their journey.[113]

Steamboats were doing a successful business on many American rivers long before the first locomotive engines pulled passenger cars. Unlike watercraft, locomotives required miles of expensive tracks and available land on which to run them. Like the Mississippi steamboat companies, the first American railroad company, the Baltimore and Ohio, sought to open western transportation, especially for the flour trade. As was true of many other private ventures using new technology, the B&O promoted itself as a company performing a public service.[114] Both federal and state governments quickly realized that railroads were indeed a public good; several states granted right of eminent domain and monopolies to railroad companies. A few states, such as Pennsylvania, owned their own railroads. The demand for railway lines became so great that the federal government granted reduced tariff fees on imported iron track.[115] Rail travel, like steam travel before it, was perceived as a driver of national progress. The *Baltimore American* told readers that the railroad was "bringing an empire to our doors."[116] The *Baltimore Patriot and Mercantile Advertiser* called the B&O "our great Railroad to the West."[117]

Enthusiasm for the railroad found expression in a song composed for the Baltimore and Ohio's opening ceremony:

> O we're all full of life, fun and jollity,
> We're all crazy here in Baltimore.

> Here's a road to be made
> With the pick and spade,
> 'Tis to reach to Ohio, for the benefit of trade.[118]

To drive home the connection between private enterprise and national good, the ceremony was held on July 4, 1828. The guest of honor was an icon of the American Revolution: ninety-one-year-old Charles Carroll, the last surviving signer of the Declaration of Independence. Music composed to honor the occasion was titled "The Carrollton March."[119]

Railroads and steam power were not inevitably linked. The Baltimore and Ohio's first railcars were pulled by horses. Only in 1830 did the company make the transition from horse power to steam power. Not everyone was convinced that steam engines were better than horses. To prove that they were, the New York engine designer Peter Cooper constructed the *Tom Thumb*, a small engine, to pull several B&O railcars. To emphasize his point that steam power was faster than horse power, Cooper arranged for a horse-drawn coach belonging to the Stockton and Stokes passenger stage company to race against the *Tom Thumb*. Benjamin Henry Latrobe Jr., an architectural engineer for the B&O, was an eyewitness to the race that day:

> At first the gray had the best of it, for his steam would be applied to the greatest advantage on the instant, while the engine had to wait until the rotation of the wheels set the blower to work. The horse was perhaps a quarter of a mile ahead, when the safety-valve of the engine lifted, and the thin blue vapor issuing from it showed an excess of steam. The blower whistled, the steam blew off in vapory clouds, the pace increased, the passengers shouted, the engine gained on the horse, soon it lapped him the silk was plied—the race was neck and neck, nose and nose then the engine passed the horse, and a great hurrah hailed the victory. But it was not repeated, for just at this time, when the gray master was about giving up, the band which drove the pulley, which moved the blower, slipped from the drum, the safety-valve ceased to scream, and the engine, for want of breath, began to wheeze and pant. In vain Mr. Cooper, who was his own engineer and fireman, lacerated his hands in attempting to replace the band upon the wheel; in vain he tried to urge the fire with light wood: the horse gained on the machine and passed it, and, although the band was presently replaced, and steam again did its best, the horse was too far ahead to be overtaken, and came in the winner of the race.[120]

It was a Pyrrhic victory for the Stockton and Stokes horse. Steam power quickly replaced animal power as new railroad lines connected towns and ports throughout the Eastern United States with inland waterways. To emphasize the eclipse of equine transportation, the illustrated masthead for the *American Railroad Journal and Advocate of Internal Improvements* showed the locomotive *Philadelphia* pulling a flatbed car with a horse carriage (sans horse) filled with gaily waving passengers.[121] Just as poets had celebrated the triumph of the steamboat a decade earlier, now the locomotive was the darling of technology enthusiasts. B&O Railway dinner plates competed with steamboat ware.

FIGURE 3.2.   "The Baltimore & Ohio Rail Road" plate. Victoria and Albert Museum, London.

The poem "Locomotives" expressed the enthusiasm with which men and women cast off their dependence on horses that quickly tired for the new, ever energetic steam engine:

Hurrah! Hurrah! Away we go—
Without a spur or goad—

Our iron coursers snort and blow
Along an iron road.

Your noblest steeds of flesh and blood,
Are soon with toil o'erdone—
But wheels impelled by fire and flood,
For ever may roll on.

No load, nor length of way fatigues
Our wild unslumbering team,
A jaunt of a hundred thousand leagues,
Is baby play for steam.[122]

Perhaps because steam technology did so much, there was speculation about what more it might do in the future. Hezekiah Niles, the editor of *Niles Weekly Register*, predicted in 1827 that "the time will soon come, when a person may pass from the city of Baltimore to some point on the Ohio river,

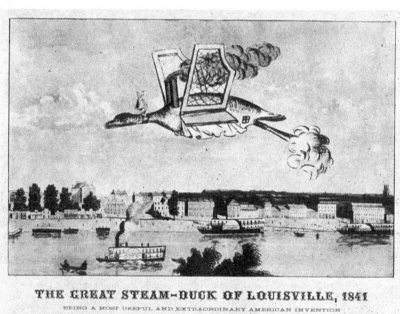

FIGURE 3.3. "The Great Steam-Duck of Louisville" (Louisville, KY: Henkle, Logan, 1841). Colonel Richard Gimbel Aeronautical History Collection, United States Air Force Academy, McDermott Library.

with the same sort of certainty, ease and convenience, that he may make a voyage from Baltimore to Norfolk in a steam boat—that little travelling palaces will be prepared, in which persons may eat, drink, sit, stand or walk, and sleep, just as they do in steam boats. Why not?"[123] Numerous satires imagined the benefits and pitfalls of technological advances. In 1841 an enterprising member of the Louisville, Kentucky, Literary Band imagined a flying machine called the Steam Duck:

> This machine, powered by steam, is calculated to revolutionize the art of aerial navigation, and to replace all aerostats, or balloons, which heretofore have proved the only possible manner of conquering the atmosphere. The Steam-Duck is fifteen feet long from beak to tail, and six feet in diameter. It is constructed in the form of a mallard duck, a fowl well known for its swiftness of wing. It is constructed of light hickory and is covered with canvas, varnished and air tight. The wings have but one joint, and are so constructed as to revolve with the necessary motion. A steam scape-pipe, passing along under the bottom, is conducted out under the tail, and gives additional impetus to the machine in its flight.[124]

One poetic fantasy of steam flight began:

> Tell John to set the kettle on,
> I mean to take a drive;
> I only want to go to Rome,
> And shall be back by five.
>
> Tell cook to dress those hummingbirds,
> I shot in Mexico;
> They've now been killed at least two days,
> They'll be *un peu trop haut*.
>
> And Tom, take you the gold leaf wings,
> And start for Spain at three;
> I want some Seville oranges,
> 'Twixt dinner time and tea.[125]

The most vivid technology fantasies of the future appeared in illustrated prints. Robert Seymour's comic "Locomotion" imagined the pleasures and pitfalls of steam-driven transportation.[126]

William Heath's "March of Intellect" depicted further outlandish steam-powered gadgets, including the "Grand Servant Superseding Apparatus," which eliminates the need for human labor.[127] Another vision of future

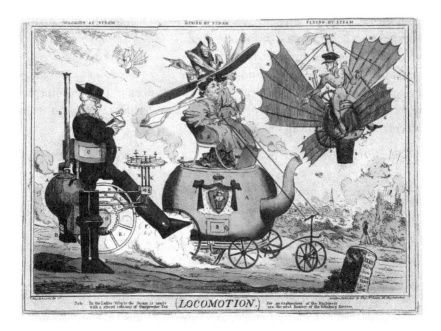

FIGURE 3.4. "Locomotion: Walking by Steam, Riding by Steam, Flying by Steam," ca. 1830. Courtesy of Lewis Walpole Library, Yale University.

improvements, with the comprehensive title "The Century of Invention Anno Domini 2000, or the March of Aerostation, Steam, Rail Roads, Moveable Houses & Perpetual Motion," was reproduced on a handkerchief.[128] Even more popular were a series of prints by G. E. Madeley depicting a Mr. Golightly riding a steam rocket. Madeley may have got the idea for his character from a patent registered in 1841 by a Charles Golightly.[129] Madeley's character first appeared in a London print, "Portrait of Mr. Golightly Experimenting on Mess. Quick and Speed's New Patent, High Pressure Steam Riding Rocket." American versions of Golightly soon followed. In 1849 Golightly made an appearance in Nathaniel Currier's "The Way They Go to California" along with Porter's steam-powered air balloon. That same year, Golightly had his own Gold Rush print, "Mr. Golightly, Bound to California." With pick and shovel strapped to his back, Golightly sits astride Quick & Speed's steam-powered rocket ("Warranted not to burst"). But Golightly is an entrepreneur, not a miner. He carries a "Patent Gold Washing Machine" and boxes of pills and tobacco; he plans to "let others do the diggings while I do the swappins!"

A less lighthearted imagining of future technologies was portrayed in the short story, "Steam." It recounts the narrator's dream (provoked by gin and water before bedtime) of a dystopian future where the trees, grass, and

FIGURE 3.5. "Mr. Golightly, Bound to California," ca. 1849. Library of Congress, Prints and Photographs Division.

open spaces of his childhood village are transformed into a sterile urban landscape with "houses, factories, turnpikes and railroads" along which "monstrous machines flew with unconceivable swiftness." When the narrator observes a worker on a seventeen-story building plunge to the street, no one is alarmed. He hears passersby comment, "'Only a steam man,' said one.—'Won't cost much,' said another. 'His boiler overcharged, I suppose,' cried a third, 'the way in which all these accidents happen!' and true enough there lay a man of tin and sheet iron, weltering in hot water." The narrator learns that these "locomotive men" are everywhere. The idyllic rural cottage where he grew up has been replaced by the Grand Union Railroad Hotel in which everything is "steam, steam, nothing but steam!" Even the staff are steam-powered: "Instead of a pretty, red-lipped, rosy-cheeked chambermaid, there was an accursed machine-man, smoothing down the pillows and bolsters with mathematical precision." At the theater, he is amused by a performance of *Hamlet* by steam: "The automatons really got along wonderfully well, their speaking faculties being arranged upon the barrel-organ principle greatly improved, and they roared, and they bellowed, and strutted, and swung their arms to and fro as sensibly as many admired actors." But overall, the narrator is dismayed by the changes he observes: "All things seemed forced, unnatural, unreal," and he is glad to wake up and realize he was dreaming.[130] "Steam" may have been inspired by Jane C. Loudon's novel that appeared a few years earlier. *The Mummy!* centers on an ancient Egyptian revived by electricity in the twenty-second century. Loudon's futurescape has steam-powered devices, including automaton doctors and lawyers.[131] *The Mummy!* and "Steam," like many speculative writings on the future, blend hope and fear. A recurring theme in these commentaries is speed: steam was already moving machines and people with a rapidity unimaginable only a few decades before. But what were the consequences and trade-offs for this? Nathaniel Hawthorne saw the railroad as the harbinger of a future world in which the countryside was devoid of tranquility. Hawthorne complained, "There is the whistle of the locomotive—the long shriek, harsh, above all other harshness, for the space of a mile cannot mollify it into harmony. It tells a story of busy men, citizens, from the hot street, who have come to spend a day in a country village; men of business; in short of all unquietness; and no wonder that it gives such a startling shriek, since it brings the noisy world into the midst of our slumberous peace."[132]

The speculative fiction of the early nineteenth century posited a vanished natural world replaced by steel, concrete, and robots. The present soon caught up with the future. In 1868 a New Jersey mechanic built a working steam man. Zadoc Dederick's invention was designed to pull a cart attached to a mechanical body. Coal and water to supply the engine were stored in the torso. To complete the effect of a mechanical man, Dederick clothed the machine in a coat and placed a sheet-iron hat (which functioned as the smokestack) atop his white, enameled face. One newspaper described the steam man's human-like "cheerful countenance."[133] Dederick first demonstrated his mechanical man at a beer hall in Newark, New Jersey. For twenty-five cents, visitors watched the steam man walk.[134] Dederick and his partner, Isaac Grass, received a patent for the vehicle a few months later. The steam man was news: reprints from the Newark papers appeared as far away as New Zealand.[135] Though Dederick planned to market the steam man, no others were made. It was not agile enough (it was unclear if the steam man's legs could move efficiently or quickly), nor cheap enough (Dederick and Grass spent $2,000 to construct it) for commercial use.

Despite technical drawbacks, within months of the first demonstration of Dederick's mechanical man, a fictional steam man appeared in print. Edward S. Ellis's *The Steam Man of the Prairies* was the first science fiction dime novel.[136] Unlike Dederick's model, the fictional steam man could run sixty miles an hour and navigate rough terrain. Ellis's story was part of the emerging genre of western adventure. In this case, the adventure consisted of battling Native Americans for a mining claim. *The Steam Man of the Prairies*

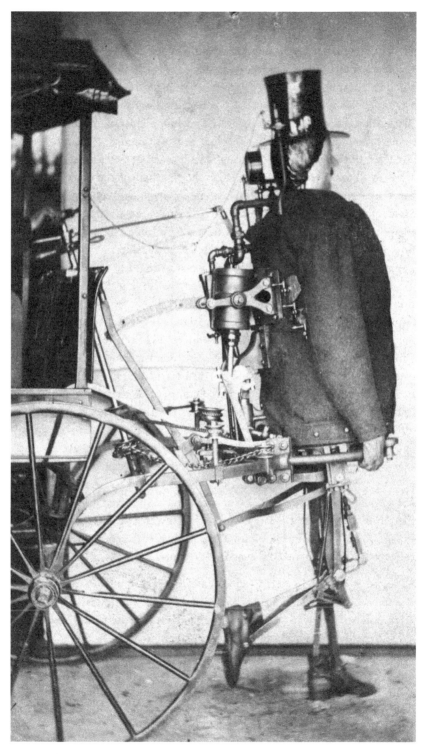

Figure 3.6. "Dederick's Steam Man," 1868. The Miriam and Ira D. Wallach Division of Art, Prints, and Photographs, New York Public Library.

was also the first in a long series of adventure stories in which technology—real or imagined—aided the hero and his companions in searching for treasure, vanquishing enemies, and eliminating Indigenous people. These stories were the wish-fulfillment of the United States' expansionist policies; just as Black Hawk was intended to be awed and subdued by air balloons, Native Americans in dime novels were defeated by steam men and steam horses.[137]

Steam travel had its proponents and its detractors. Steamboat explosions were an all-too-frequent occurrence in the first half of the nineteenth century, often making the front page of newspapers throughout the United States and overseas. As knowledgeable consumers, Americans were aware of the potential dangers of steamboat travel. An advertisement for the steamboat *Balloon*, for instance, explicitly stated that the ship used a low-pressure engine; customers knew that the high-pressure engines were more likely to explode.[138] Steamboat explosions, especially before effective federal regulation in the 1850s, were frequent and deadly.[139] They were also newsworthy. When the *Lexington* exploded in Long Island Sound in 1840, the *New York Sun* put out an extra edition to cover the details of the calamity. Nathaniel Currier issued fourteen separate images, including a color lithograph, of the event. Despite the alarming number of deaths and injuries suffered by steamboat workers and passengers, and the ubiquity of news coverage and visual imagery, this popular culture of disaster, as Cynthia Kierner has termed it, did not diminish Americans' embrace of steam travel.[140] Men and women rode steamboats for pleasure and employed them for profit. Travelers along the Eastern Seaboard rejected bumpy, cramped stagecoaches for comfortable boat cabins. Midwestern farmers shipped grain to New Orleans in a few days. And gold prospectors from eastern cities arrived in California in a matter of weeks. Despite danger, noise, and inconvenience, people admired the products of steam technology. Though Harriet Beecher Stowe preferred canal boats, she recognized the steamboat's attractions:

> There is something picturesque, nay, almost sublime, in the lordly march of your well-built, high-bred steamboat. . . . like some fabled monster of the wave, breathing fire, and making the shores resound with its deep respirations . . . there is something mysterious, even awful, in the power of steam. . . . But in a canal boat there is no power, no mystery, no danger; one cannot blow up, one cannot be drowned, unless by some special effort: one sees clearly all there is in the case—a horse, a rope, and a muddy strip of water.[141]

Stowe's observation brings us back to William Bentley and his vigilance for both sea monsters and steamboats. From 1817 until 1819, Bentley noted reports of sea serpent sightings. Bentley himself saw it only once, in September 1817.[142] Two months earlier, after many observations from the shore, Bentley had taken his first steamboat ride, aboard the *Massachusetts*, from Salem to Boston and back. Bentley wrote in his diary: "The motion was easy & regular throughout the passage. . . . The Boat answered every expectation & when an accommodation is given for landing at every tide, [I] will find this cruise a safe & pleasant one."[143] After that first ride, Bentley took two more excursions before the *Massachusetts* ceased its journeys for the winter. Bentley could rest content that he had seen both natural and man-made aquatic novelties, both "fabled monster(s) of the wave."

In the early nineteenth century, new technologies came thick and fast; steam-powered machinery, locomotives, and ships amply demonstrated that they could improve travel and increase productivity. Although perpetual motion machines were a continual disappointment, Americans' embrace of successful technologies drove the economic engine of the nation. The same year that William Bentley took his first steamboat ride in Salem, the poet Fitz-Greene Halleck celebrated the technical wonders of the age in his poem "Fanny":

> We owe the ancients something. You have read
> Their works, no doubt—at least in translation; . . .
> 'Twas their misfortune to be born too soon
> By centuries, and in the wrong place too;
> They never saw a steam-boat, or balloon,
> Velocipede, or Quarterly Review;
> Or wore a pair of Baehr's black satin breeches.
> Or read an Almanac, or C*****n's Speeches.[144]

Halleck's smug satisfaction in the nation's technological accomplishments reflected what nineteenth-century Americans were most proud of. By the 1850s, steam technology was so much a part of American life that one commentator described his fellow citizens as becoming a "living organic unit with railroads and steam-packets for a circulating system, telegraph wires for nerves, and the London *Times* and New York *Herald* for a brain."[145] With the aid of steam engines, the United States was quickly becoming the nation its citizens aspired for it to be.

CHAPTER 4

Grand Designs

Technology and Urban Planning

Americans did not simply talk about becoming a great nation, they did something about it. Mechanics' societies and schools encouraged workers to learn the skills necessary to make and invent. Contests lent incentives to develop or improve methods and machines. Speeches, essays, and poetry lauded Americans' ingenuity and enterprise. The United States saw itself as an emerging empire; western expansion and settlement, an assertion of authority in the American hemisphere, a strengthening economy, and increased military power all contributed to the nation's standing on the world stage. Water projects exemplified these national ambitions: they were large, expensive, and visible. They affected the lives of thousands of people every day. Their physical presence, often in the center of cities, made them sites for recreation and tourism. Visually, these functional buildings were designed and embellished as echoes of ancient empires. Born out of necessity, water systems were promoted, celebrated, and identified as representations of America's place among the nations of the world and as the inheritor of past civilizations.

The Water Problem

The monumental reservoirs built in the early nineteenth century were intended to solve two perennial challenges for urban dwellers: fire and disease.

Boston and New York experienced significant losses from fires in the eighteenth and early nineteenth centuries. The first Great Boston Fire occurred in October 1711. Cotton Mather used the awful occasion to remind Bostonians of God's judgment on unrepentant sinners.[1] The second Great Fire of Boston, in March 1760, destroyed 349 buildings, including the Quaker Meeting House and several ships docked at Long Wharf. This fire, too, inspired a sermon. The Congregationalist minister Jonathan Mayhew published "God's Hand and Providence to be religiously acknowledged in public Calamities: A Sermon Occasioned by the Great Fire in Boston, New England."[2] Governor Thomas Pownall called upon the inhabitants of the Massachusetts Bay Colony to contribute to the financial relief of the 220 families left homeless by the conflagration.[3] Despite a subsequent law requiring new buildings to be brick rather than timber and the establishment of an official fire department, homes, shops, warehouses, museums, Faneuil Hall, the Exchange Coffee House, and even a portion of the State House were all ravaged by fires.

New York City fared no better than Boston. During the British occupation in 1776, fire destroyed one-quarter of the city.[4] An arson panic seized the eastern coastal cities in 1803. Newspapers fueled the alarm with reports of houses deliberately set alight. Elizabeth Drinker of Philadelphia copied the notices into her diary verbatim and then wrote: "It is supposed by many, that the fire at Portsmouth was done by design—and one lately in New York. How dreadfully wicked must the perpetrators be!" A day later she copied from *Bronson's Evening Paper* a notice that arsonists had attempted to set fire to a stable in Market Street: "A log of dry oak, lighted, inclosed in a combustible wrapper, was thrown among the straw; and had not the merest accident frustrated the nefarious project, the whole place must have been destroy'd."[5]

City dwellers were not without resources to fight fires. Fire engines were introduced as early as 1731. These boxes on wheels enabled water to be pumped through a hose, providing enough force for a heavy stream of water. But eighteenth-century technology was only as good as the water supply. Firefighters used leather buckets to carry water from wells and public pumps to fill the engine's tank. Without a reliable water source where and when it was needed, buildings were at the mercy of happenstance.

Urban dwellers required clean as well as accessible water. Public pumps and private wells supplied most residents. A few houses had rooftop cisterns. Some New Yorkers purchased water; it was delivered to them in barrels.[6] As cities became more crowded, private wells proved insufficient for the needs of more and more inhabitants. And wells, whether private or public, were often dangerously close to privies, a connection that did not go unnoticed. Henry J. Latrobe, in his 1798 engineering report to the city of

Philadelphia on the options for constructing a water system noted that "the perfect permeability of this stratum is evident from the connection of the wells with each other, and with the sinks and privies, from whence arises the extreme unpleasantness of the water in the crouded parts of the city."[7] Privately, Latrobe was more strident in his criticism of Philadelphia's water: "The Water is not to be drank, and it is worst in the crouded neighbor-hoods . . . [tasted] as if it contained putrid matter."[8] Latrobe's observation was echoed five months later in the Philadelphia *Aurora*. The author decried the "baneful custom" of digging privies only twenty or thirty feet deep—shallow enough that "by the means of the absorbent powers of the gravel that is generally found at a certain distance beneath the surface of the earth, they must naturally communicate their excrementious qualities to the waters of the city."[9] The most serious public health concerns connected with the water supply were the outbreaks of yellow fever between the 1790s and the 1830s. Latrobe speculated that water pumps channeled "volumes of noxious Gas from the putrifying water" that, if not a direct cause of yellow fever, might have been a contributing factor. The real cause of yellow fever, infection carried by mosquitoes, was unknown until the twentieth century. Eighteenth- and nineteenth-century physicians blamed contaminated air—a miasma of "putrid exhalations" caused by the filthy conditions in urban streets.[10] The solution was routine cleansing of the streets to eliminate the cause of the fever. Disease outbreaks in all the eastern port cities between 1791 and 1799 created an urgency to harness a reliable and copious water supply.[11] When cholera epidemics supplanted yellow fever in the early decades of the nineteenth century, "putrid exhalations" were still held to blame. The urgency for a clean and plentiful water supply intensified. Citizens were confident that well water was safe to drink, but there was a need for much more water than the street pumps could supply in order to keep the dust, dirt, and fumes in check. When, during the worst cholera outbreaks in the 1830s, Philadelphia suffered fewer deaths than other cities, credit went to continuous street cleaning. The water that scoured the streets came from the Latrobe-designed system, which eliminated the necessity of drawing from wells and cisterns and guaranteed cleaner water.[12]

Social concerns dovetailed with health concerns: as reform-minded citizens advocated for temperance in the 1830s, water was the alternative to alcohol. Some believed that bad water was responsible for men and women turning to drink.[13] Availability was key, but so was taste. As Latrobe noted of Philadelphia's noxious-tasting water, many urban dwellers avoided consuming water because of its unpleasant taste or smell, or both. In Boston, Loammi Baldwin Jr., the engineer responsible for designing the city's public water system, argued that the "free and copious introduction of pure, soft water" would help "to expunge from the community a loathsome indulgence in vitiating liquors."[14] At the ceremony for the opening of the Boston water system, the mayor Josiah Quincy Jr. commented that the construction of the system had been accomplished "without the stimulus of intoxicating liquor."[15] Unsurprisingly, no alcohol was served at the festivities. In contrast, the budget for Philadelphia's water system included supplying the workers with alcohol.

The Water Solution

After decades of yellow fever outbreaks, the devastating epidemic in the summer of 1798 prompted the city of Philadelphia to act. A petition to the city's Select and Common Councils resulted in the creation of what became known as the Watering Committee.[16] The committee hired the engineer Benjamin Latrobe to study the problem of how to deliver water to the streets and homes in the city. Latrobe's published report outlined several alternatives and costs.[17] After much complaining and political maneuvering, the city agreed to finance the water system with a special tax. It took twenty months, from 1799 to 1801, and $220,000 (including nearly $1,000 for alcohol).[18] Philadelphia's water system was less than democratic: the municipality assumed the costs of laying pipes from the pump station at Centre Square through the city streets. Wealthy property owners paid to have pipes connected to their homes, while the poor used public hydrants.[19] Latrobe calculated the price of connecting to the city water system to be fifty cents per foot. On average, a homeowner would pay twenty-five to thirty dollars, an expense, Latrobe confidently assumed, "which I think every family would cheerfully incur to avoid the inconveniences arising from the necessity, as at present, of sending their servants to the pumps."[20]

Latrobe's design solved the problems of both supply and demand. Steam-driven pumps on the Schuylkill River at the foot of Chestnut Street conveyed water to a tank in Centre Square (at the intersection of Broad and Market). From there, water pipes laid throughout the city went into private homes and street corner hydrants. Finally, there was water enough to drink, to clean the streets, and to put out fires. The *Philadelphia Gazette and Daily Advertiser* assured readers that Schuylkill water was safe to drink: "There can be no doubt but that for drinking and culinary purposes it will be found infinitely preferable to the water of our pumps."[21] And citizens' homes were now protected from fire: the pipes installed under the streets had valves to allow water to be diverted into hoses. Latrobe considered fire control alone worth the price of the water system.

The Philadelphia water system was a technological triumph over dirt, destruction, and disease. One celebratory poem praised the waterworks for its multiple benefits:

The Western acid, and the Eastern sweet,
With *l'eau de vie* shall in the Schuylkill meet
To give cold comfort in the cool retreat.
Brimful of kindness to the cleanly wishes
Of cooks or scullions washing greasy dishes,
To bake, to boil, to wash, or brew how handy,
'Tis worth a sea of whiskey, gin or brandy.
Kind humble stranger wash our woes away,
Filth, fish-guts, fevers, wash them to the sea;
When fire is cry'd the ladies' fears assuage,
Stop its dire progress and o'erwhelm its rage.
When will thy sickness cease, when shall we meet
Thy boasted blessing in each happy street?
Or will thy bursting pipes continue spuing,
Or Hydrothorax prove thy utter ruin?
Dear WATER-WORK's cheap stock, dear loan
Do thou support without a sigh or groan;
With fuel let the ENGINES be well fed,
"And on the waters cast thy daily bread."[22]

Architecturally, the waterworks were intentionally grand. Latrobe's design drew attention to Philadelphia's link with ancient empires. The Centre Square engine house was an elegant, marble-clad building that echoed the monuments of ancient Rome. In his speech to the Society of Artists, Latrobe invoked the classicism that inspired the nation's architects to emulate the ancients: "The days of Greece may be revived in the woods of America, and Philadelphia become the Athens of the Western world."[23] Latrobe's ambition for the city was a result of the culture of classicism that infused literary, political, and cultural life in the early United States. Americans looked to Greece and Rome as models for political ideals and civic culture. The revolutionary generation's imagined affinity with the ancient republics meant that Americans drew on the heroic past to create a modern polity; the Roman Senate represented liberty, and the soldier farmers of Sparta were "suitable exemplars for America's young agrarian republicans."[24] This admiration for ancient empires was expressed through classical allusions, classical images, and neoclassical buildings. The citizen soldiers of the American Revolution named the Society of Cincinnati after the Roman who led his country through war and then left public office to return to his farm. These same veterans christened newly surveyed territory with classical names such as Syracuse, Cicero, Tully, Romulus, Utica, Pompey, Ithaca, Marathon, and Cincinnatus, emphasizing that a new republic was in the making. Latrobe's design for Centre Square conveyed the importance of this technological triumph to the health, safety, and prosperity of the city by visually echoing the monuments of the ancient world.

Centre Square soon became a popular gathering place for recreation and celebration. The artist Lewis Krimmel captured the activities and amusements that Philadelphians pursued there in his 1812 painting *Fourth of July in Centre Square, 1812*. Krimmel's image highlights the large fountain, ornamented with a sculpture of a Roman water nymph, that advertised the triumph of the city's water system.[25] Philadelphians celebrated the anniversary of independence by gathering at a monument to republicanism.

Even the ornamental had a practical side. Fountains cooled and cleansed the air. This was an important consideration for those who believed that bad air was the cause of disease.[26] Capitalizing on the popularity of Centre Square, George Blake illustrated his *Collection of Duetts for Two Flutes*,

FIGURE 4.1.   John Lewis Krimmel, *Fourth of July in Centre Square* (1812), Pennsylvania Academy of the Fine Arts.

Clarinets, or Violins with an image of the building.[27] Long after it was demolished in 1829, Centre Square reservoir still resonated with Philadelphians as a place for recreation and a symbol of neoclassical elegance. The *Casket*, a monthly magazine, included an image of the building and grounds, showing Philadelphians enjoying a stroll in front of William Rush's *Water Nymph*.[28] As late as 1860, the *Philadelphia Ledger* gave an illustration of the reservoir as a New Year's gift to subscribers.[29]

By the early 1820s, a new reservoir on Fairmount Hill, powered by waterwheels rather than a steam engine, was housed in another neoclassical building. Here, temples and porticoes echoed Greece rather than Rome.[30] Like Centre Square, Fairmount Hill was intended to be an exemplar of technological achievement in the new republic, and its architecture and ornamentation conveyed grandeur and power. The Fairmount Water Works were designed to be seen and to be celebrated. The sculptor William Rush again produced art suited to the scene. His pair of figures, *Allegory of the Schuylkill River in Its Improved State* and *Allegory of the Waterworks*, suggested that only with the aid of man could nature be truly useful. *Allegory of the Schuylkill River in Its Improved State* is the river embodied by a chained man, a river god under the control of the dams and locks that harness the river's power.[31] *Allegory of the Waterworks* is a graceful Grecian woman leaning on a fountain and casually reaching toward the wheel that powers the waterworks pump.

Visitors praised the waterworks. As John Sheldon of New York recounted to his wife, while in Philadelphia he had "visited almost every place of note." This included the waterworks, a technological triumph he described as "beyond all praise." He observed that the system not only supplied ample water to the city, but equally important, "the water of the Schuylkill is of an excellent quality, and as you have, in the city of New York, often felt the importance of good water, you can readily imagine the luxury in this respect, which is enjoyed by the inhabitants of Philadelphia."[32] The inventor Thomas Ewbank bestowed the Philadelphia engineering marvel with the supreme accolade: "It is impossible to examine these works without paying homage to the science and skill displayed in their design and execution; in these respects no hydraulic works in the Union can compete, nor do we believe they are excelled by any in the world. . . . The picturesque location, the neatness that reigns in the buildings, the walks around the reservoirs and the grounds at large, with the beauty of the surrounding scenery, render the name of this place singularly appropriate."[33]

Groups of visitors made a point of including the waterworks in their itinerary, as when Captain Latrobe's Light Infantry of Maryland visited the city in 1827. In addition to entertainment at theaters and civic dinners, the men took in the Fairmount Water Works.[34] When Chief Blackhawk and his fellow captives were taken to Philadelphia in 1833, they were shown the Fairmount Water Works, the United States Mint, and the newly built Cherry Hill prison.[35] All were symbols of American supremacy. Guidebooks to the city included descriptions of the waterworks. Some, such as James Mease's *Picture of Philadelphia*, highlighted the buildings as one of the most important spots to visit in the city. The *Port Folio* chose an illustration of the waterworks for its cover in 1819. The publishers Carey and Lea used an engraving of Thomas Doughty's painting *Fairmount Waterworks* (1822) as the frontispiece for their guidebook, *Philadelphia in 1824*.[36] *The Stranger's Guide* (1828) underscored the importance of the waterworks as a tourist spot by reproducing Thomas Birch's "The Water Works at Fair-Mount, Philadelphia," as the *Guide*'s only illustration. In the foreground, Birch placed a steamboat in the river, thus celebrating two triumphs of technology made possible by the steam engine.[37]

By 1849 guidebooks included omnibus schedules from the center of town out to the Schuylkill. George Appleton's *A Handbook for the Stranger in Philadelphia* informed readers that there were "numerous omnibuses running to this beautiful and romantic spot; they leave the Exchange every fifteen minutes, and after a pleasant quarter of an hour's ride through Chestnut and Broad and the intervening streets, arrive at Fairmount."[38] Eli Bowen's *The Pictorial Sketch-Book of Pennsylvania* assured readers that the Philadelphia waterworks "are eminently worth a visit from a stranger."[39] As a result of the waterworks' popularity, the engine house became a public saloon where visitors enjoyed refreshments. The mill house was renovated to provide visitors a view of the water wheels.[40] Appleton directed his readers to the heart of

FIGURE 4.2.   Thomas Birch, *View of the Dam and Water Works at Fairmount, Philadelphia* (1824), engraved by Robert Campbell (Philadelphia: Edwd. Parker, 1824). Pennsylvania Academy of Fine Arts.

THE

PORT FOLIO.

VOL. 7.

——1819——

VIEW OF FAIR MOUNT WATER WORKS PHILAD.ᵃ

Published by Harrison Hall

Nᵒ. 209 Chesnut Sᵗ.

PHILADELPHIA.

FIGURE 4.3. The *Port Folio* (1819). Author's collection.

the waterworks in the mill house, where the giant machinery pumped river water up to the reservoir and out to the city.[41] Appleton devoted several pages to the history of the waterworks' construction and provided statistics on gallons produced per day (530,000), gallons consumed per day (4,000,000), and the cost of the water to city residents (four dollars per year). Readers were left in no doubt about the importance of this technological achievement.[42]

Foreign as well as domestic visitors put the Fairmount Water Works on their itineraries. The Duke of Saxe-Weimar described his visit in his published account of his travels in the United States.[43] Frances Trollope, who made a point of going everywhere and seeing everything, noted in 1830 that the popularity of the waterworks was such that "several evening stages run from Philadelphia to Fair Mount for their accommodation."[44] Ten years later, Charles Dickens praised the waterworks for supplying the entire city of Philadelphia, "to the top stories of the houses," for "a very trifling expense."[45] While working as a typesetter at the *Philadelphia Inquirer*, Mark Twain described his journey on the Fairmount stage to view the waterworks. He was especially taken with Rush's *Water Nymph*, "the prettiest fountain I have seen lately. A nice half-inch jet of water is thrown straight up ten or twelve feet, and descends in a shower all over the fair water spirit. Fountains also gush out of the rock at her feet in every direction."[46] Twain was not alone in his admiration of the waterworks. One magazine commented in 1832 on the pleasure the site conveyed: "Crowds of visitors increase every day, while their gratified faces tell very plainly that they are viewing the most gratifying sight in America."[47] These same crowds may have continued their pleasure as they danced in their parlors. "The Fairmount Quadrilles," published in 1836, were lively square dances. The sheet music, illustrated with a view of the waterworks, was the souvenir of a pleasurable afternoon.[48]

The popularity of the waterworks reflected how people spent their leisure time and the value they placed on new technologies that enhanced their lives. Admiration for these modern wonders continued beyond a Sunday's excursion. The Englishman George Brewer's panorama, a popular visual entertainment in the early nineteenth century, showed audiences this important structure. Itinerants such as Brewer took their large canvases from city to city, or even country to country, giving armchair travelers the pleasure of sights without the trouble of travel. Between 1848 and 1850, Brewer exhibited his panorama in Boston, Cincinnati, Louisville, Philadelphia, New Orleans, and St. Louis. He chose several of the most remarkable natural and man-made sights for his panorama: Mammoth Cave in Kentucky, Niagara Falls, the western prairies, and the Fairmount Water Works. All of these, Brewer stated, were "American wonders." As "great national objects, they have been selected for this truly national production."[49] Brewer told his audience: "Fairmount furnishes one of the most beautiful and useful combinations of nature and art to be seen in the whole country."[50] Fairmount was a triumph of man harnessing nature to suit his purposes.

The waterworks drew artists as well as sightseers. Fairmount was the subject of paintings, lithographs, and engraved illustrations in popular magazines. At least seventeen images of the area had been published by the 1830s.[51] The views combined man-made ingenuity with the site's natural beauty. Frederick Graff designed a neoclassical shell to house the waterworks' machines, and he landscaped the surrounding acres to complement his engineering marvel. Visitors admired the waterworks, walked its veranda, and strolled along paths lined with statues (including Rush's *Water Nymph* relocated from Centre Square) to the well-situated pavilion overlooking the Schuylkill dam. Prints re-created this simultaneous experience of nature and technology.[52]

Perhaps the most telling signs of how technology resonated in nineteenth-century American life are the images of bridges, buildings, and water systems depicted on pots, bowls, and plates designed for domestic use. Americans served tea from teapots illustrated with the Fairmount Water Works. Vases illustrating the buildings stood in halls and parlors. Meals were served on illustrated plates and bowls. Much American dinnerware was imported from England, but canny manufacturers, eager to capture the American market, catered to American tastes. Most popular among export wares were American views—especially significant landmarks. Joseph Stubbs used an engraving of Thomas Birch's popular illustration, *Fair Mount near Philadelphia*, to decorate a complete earthenware dinner service sold in the 1830s.[53] It was fitting that the makers of porcelain vases, the Tucker and Hemphill china factory, stood at the foot of Chestnut Street, on the site of the first pumping station for Centre Square. Tucker and Hemphill produced two vases, almost two feet tall, with paintings of the waterworks. Executed in neoclassical design, they shine with gilt decorative touches and brass griffins for handles. Clearly meant to be conversation pieces, these elaborate and costly domestic embellishments showcased the pride Philadelphians expressed for this modern technology.[54] The Fairmount Water Works were many things— a water supply, a destination, and a symbol (embellished on a water vase)— of American enterprise.

"A Great Act in the Mighty Drama of a Nation's History"

Other cities took note of Philadelphia's technological triumph. The municipal leaders of Boston, New York, and New Orleans studied ways and means for their cities to obtain an adequate water supply for fire prevention and for domestic and industrial use. Bostonians observed that Philadelphia's fire prevention system, with hydrants located on almost every block of the city, gave pumping engines ready access to a continuous supply of water. Mayor Quincy wrote to the chairman of the Philadelphia Watering Committee to seek advice. Quincy hoped to abandon Boston's inadequate and antiquated "system of forming lines of citizens & passing buckets at fires."[55] But this could only be done if there was a sufficient and reliable supply of water at public hydrants. City Boards of Health also noticed that Philadelphia's mortality rates from the 1832 cholera epidemic were far lower than cities without public water systems. Philadelphia reported nine hundred deaths from cholera, whereas New York had thirty-five hundred and New Orleans almost five thousand.[56]

But it was New York City, not Boston, that next followed Philadelphia's lead. New York's first citywide water supply was built by private enterprise. Unlike Philadelphia, the Manhattan Company sourced its water from a well within the city. From the well, a steam engine pumped the water to a reservoir on Chambers Street. From there, a system of wood pipes distributed the water to subscribers throughout Manhattan.[57] Everyone else still relied on private or city-owned wells. The Manhattan Company was out to impress (and recruit) customers. Just as Philadelphians celebrated the technological triumph of their water system with evocations of the ancient world, the Manhattan Company adorned the most tangible source of its water in a similar manner: the Chambers Street reservoir was designed as a Roman temple, with a statue of the water god, *Oceanus*, surmounting the entrance.[58]

Though not as imposing as the Fairmount Water Works buildings and statues, the Chambers Street edifice nevertheless conveyed the importance of the building for the city's water supply, and evoked associations with the technologies of ancient empires. Though the Manhattan Company's water system was a great improvement over individual wells, the city's water usage quickly outstripped the company's ability to supply customers. New York's rapid growth in the early nineteenth century placed a strain on its water supply, a burden that increasingly could not be met by the Manhattan Company's system. In 1842 a new municipal water system, this time supplied by the Croton Aqueduct, brought enough water to meet the city's demands.[59]

And as in many other things, New York celebrated its new water system on a grand scale. On October 14, 1842, thousands of spectators turned out to see the largest procession ever staged in the city. City and state dignitaries, foreign consuls, judges, lawyers, clergymen, professors, merchants, tradesmen, artisans, soldiers, and sailors all marched from Broadway to City Hall

FIGURE 4.4. George Hayward, *Reservoir of Manhattan Waterworks. Chamber Street, 1825* (1855). New York Public Library.

Park. The opening day ceremony emphasized how important the water system was to the city. Moreover, the speeches given, the groups who participated, and the large number of spectators, demonstrated how the New York community not only valued water, but appreciated—and deliberately celebrated—the technology that made the system possible. Technology brought the city into the modern era: it enabled better health, cleaner streets, more industry, and the promotion of social reforms.

The civic procession through the streets of New York re-created the community and represented the areas of life that the Croton Aqueduct improved: fire prevention, health, and industry. The city's printers, assembled on a carriage that bore a printing press used by Benjamin Franklin, tossed copies of the "Croton Aqueduct Ode" to the crowd.[60] Fire companies, with their elaborately painted engines and banners, were present in large numbers. Images and symbols on the hose engines emphasized water's power to protect citizens from harm. The carriage of the Croton Hose Company, for example, was painted with a view of Genesee Falls on its front and Neptune and Amphitrite on its back. Several of the marchers carried the Croton banner, showing a human embodiment of the aqueduct "presenting a goblet of the water to the Queen of Cities, who is crowning a fireman with a wreath." Another figure, the "Fire King," lay chained at her feet, while "Neptune stands quietly looking on, and Manhattan is about retiring, as his services are no longer required." The Columbian Hose Company's engine depicted a woman fleeing from a burning building with a child in her arms and a view of Niagara Falls.[61] Water as an antidote to alcohol also played a prominent role: numerous temperance societies marched in the procession. The Union Hook and Ladder Company's engine displayed commitment to the cause. It bore the motto: "We are pledged to abstain from all intoxicating drinks."[62] The technology that made the aqueduct possible was also on display: the masons, contractors, and workmen who built the aqueduct marched alongside a section of water pipe drawn on a cart, while members of the Mechanics' Institute displayed a working miniature steam engine.[63]

When the procession arrived at City Hall Park, the centerpiece of the water system, the Croton Fountain, was turned on for the first time. The largest fountain in the United States, it served as a visible reminder of the city's achievement. One observer expressed his appreciation of this tangible symbol of accomplishment: "Its magnificence does not consist in its artistic features: these are very simple. It is the size and height of its central jet: the extreme beauty of its numerous arching jets, when in full play, exhibiting in the sunlight all the gorgeous tints of the rainbow."[64] Another boasted, "Its copiousness of waters is so great, that two of its fountains daily throw away more water, than suffices for the supply of other large cities."[65]

FIGURE 4.5. "Croton Water Celebration," Sidney Pearson (music) and George Pope Morris (lyrics) (New York: J. F. Atwill, 1842). Frances G. Spencer Collection of American Popular Sheet Music, Baylor University Library.

Author Lydia Maria Child thought the fountain surpassed any in Europe. She later praised the fountain's benefits to the poor of the city in her poem, "The New-York Boy's Song":

Poor little ragged children,
Who sleep in wretched places,
Come out for Croton water,
To wash their dirty faces.[66]

Images of the Croton Fountain were everywhere. A lithograph of the fountain illustrated the title page for the "Croton Jubilee Quickstep."[67] Plumbers offered customers access to the new water system and advertised bathtubs, kitchen sinks, and ornamental fountains. Merchants used the Croton's fame as a marketing ploy: the pen maker Joseph Gillott, for example, advertised the Croton pen. Presumably, its ink flowed as smoothly and copiously as water in the aqueduct. The plumber Thomas Dusenbury's trade card, illustrated with the fountain, simply stated "Croton Water" amid images of his wares and services. The word *Croton* was shorthand that conveyed a world of meaning to Dusenbury's customers: clean, abundant water reliably conveyed to urban homes.[68]

The distributing reservoir at Murray Hill was a key component of the Croton water system. Located on the west side of Fifth Avenue, between Fortieth and Forty-Third Streets, it occupied three city blocks. The forty-four-foot-high Egyptian Revival structure visually expressed the technological wonder of the aqueduct and announced to the world the arrival of the United States as a rising empire.[69] The break with neoclassical style was not as abrupt as it might seem. Americans continued to look to the classical world for political ideals and civic culture. Latrobe's Centre Square reservoir and Graff's Fairmount buildings were part of this long-established tradition. But Greece and Rome were not the only ancient civilizations from which Americans drew inspiration. Geopolitical events in the Atlantic world at the end of the eighteenth century acquainted Americans with Egyptian history and culture. The advent of Egyptian style in the United States was directly linked to Napoleon's invasion of Egypt in 1798. Although the military consequences of the occupation were disastrous for the French, the scientists and artists who accompanied the expedition published texts and images that stimulated interest in ancient Egypt. An English translation of Baron Vivant Denon's book *Travels in Upper and Lower Egypt during the Campaigns of General Bonaparte* was for sale in New York by 1803. Denon's work contains detailed drawings of towns, tombs, and temples. For many Americans, Denon's images were their first look at Egypt.[70] The initial volume of the expedition's official publication, *Description de l'Égypte*, appeared in 1809. Both works transmitted information about ancient Egypt to Europe and America.

For most Americans, their visual acquaintance with the Land of the Pharaohs was in the form of entertainment: panoramas, museum collections, and, most exotic of all, mummies. Great cities, dramatic battles, and exotic locales were illustrated in spectacular fashion. In 1809 viewers in Charleston, South Carolina, saw a view of the Nile at the Cosmorama on Meeting Street.[71] At the celebration of peace between America and Britain in February 1815, the Temple of Peace erected on a stage at City Hall included an Egyptian Pyramid.[72] That same month Scudder's Museum offered New Yorkers "a grand view of the banks of the Nile," with Cairo and the pyramids at the Optical Cosmorama.[73] More than twenty years later, a "Description of a View of the Great Temple of Karnak and the surrounding City of Thebes" was exhibited at the Panorama building on Broadway.[74] In 1849 George Robins Gliddon exhibited a panorama of "The Nile, through Egypt and Nubia," to accompany his travel lectures in New York.[75]

Given the abundance of sources from which Americans learned about ancient Egypt, it is not difficult to understand why the Land of the Pharaohs caught the imagination of those who believed the United States was another great empire. The Mississippi River was referred to as the "American Nile," and new communities along the river were christened Karnak, Thebes, and Memphis. Planners hired the architect William Strickland to design the entire town of Cairo, Illinois, with an Egyptian obelisk as its prominent feature.[76] Nineteenth-century interpretations of the culture of ancient Egypt made Egyptian Revival style an appropriate choice for public structures. The notion of permanence and solidity it conveyed reassured a cautious public uncertain about new technologies such as steam engines and suspension bridges. A number of railway stations displayed Egyptian columns and motifs. Bridge supports echoed the sturdy pylons of Egyptian temples. Egypt was thought of as the land of ancient secrets and knowledge, therefore the wisdom and mystery associated with Egyptian architecture was appropriate for fraternal lodges, libraries, and medical colleges.

The monumentality of Egyptian architectural style was also employed to illustrate institutional authority. Just as the interiors of early nineteenth-century prisons reflected a new approach to discipline and punishment, the facades of these buildings were deliberately crafted to convey an unbreachable solidity. John Haviland designed a Gothic fortress for the exterior of the Eastern State Penitentiary in Philadelphia. But when the facade was criticized as too gloomy, too reminiscent of a dungeon, Haviland chose Egyptian style for his next prison commission, the New Jersey State Penitentiary in

Trenton (1832), and again for the Essex County courthouse in 1836.[77] When Haviland chose a neo-Egyptian design for these projects, he required an alternative to Gothic style that his audience would "read" correctly. In other words, Haviland's Egyptian Revival buildings are testimony to the familiarity of Egyptian style; Americans interpreted these buildings correctly because they knew, after reading books, attending lectures, seeing mummies and paintings of Egyptian ruins, what Pharaonic design was meant to convey. On the other hand, it is not an accident that there are almost no examples of Egyptian style in American domestic architecture. This usable past was deliberately chosen for its representation of power, authority, and stability—ideals that nineteenth-century designers put into the brick and mortar of public structures. Homeowners, unless they aimed to be domestic tyrants, sought less severe styles.

Thus, by the time the Murray Hill reservoir was completed, New Yorkers already had a reference for Egyptian Revival architecture in public buildings. The reservoir was built three years after the city's Halls of Justice, or the "Tombs," as they were commonly referred to because of their Egyptian style. The first view of the reservoir, which for many New Yorkers was the illustration on the front page of the *Dollar Weekly* for October 22, 1842, would not have disconcerted them. Constructed out of necessity, the reservoir quickly became a place of recreation and a tourist destination. Guidebooks described the structure in detail and informed readers that from the walkway along the top of the building they would see "the whole upper portion of the city and surrounding scenery. Access to this noble promenade is free."[78]

City reservoirs constructed soon after Murray Hill were also in the Egyptian Revival style. In Philadelphia, the Spring Garden waterworks supplied water to sections of Philadelphia not served by the Fairmount water supply. Visitors entered through a grand doorway reminiscent of an Egyptian temple. The chimney stacks were disguised as temple columns embellished with lotus leaves. In Albany, New York, the city reservoir was enclosed by walls in the Egyptian style.

Though advertised as a place with panoramic city views, the Murray Hill reservoir was designed to be the visual expression of what New Yorkers' believed the Croton Aqueduct had achieved. In congratulating the citizens of New York, President Tyler said the aqueduct was "justly to be classed among the first works of the age and is honorable to the enterprise of the great centre of American trade and commerce."[79] The British consul went even further in placing the aqueduct "among the greatest enterprises of any nation on earth, governed by, and voluntarily paid for, by the people. Tyrants have left monuments which call forth admiration, but no work of a free

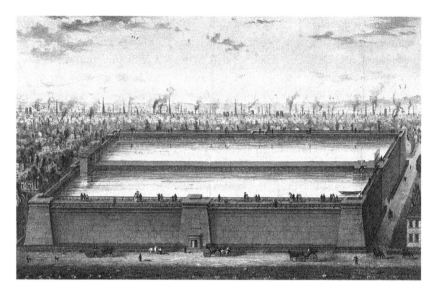

FIGURE 4.6.   N. Currier, *View of the Distributing Reservoir: On Murrays Hill—City of New York* (1842). Library of Congress, Prints and Photographs Division.

people, for magnitude and utility, equals this great enterprise."[80] The Joint Committee on the Croton Aqueduct agreed. As their report stated: "There is not an instance on record in which the citizens of any country have, of their own free will and accord, authorized the construction of a work of the same magnitude, the beneficial effects of which will be experienced by ages yet unborn." The committee believed the aqueduct would "vie in magnitude with any in the world."[81] Governor William Seward called the aqueduct "a great act in the mighty drama of a nation's history."[82]

Six years after the Croton celebrations, Boston completed its own municipal aqueduct. As in New York, the precursor to a public water supply was a private water company. The Boston Aqueduct Corporation, begun in 1798, pumped water from Jamaica Pond in Roxbury and conveyed it through pipes to subscribers' homes.[83] In other parts of the city, such as the Broad Street neighborhood near the wharves, homeowners paid six dollars a year for water from a private well.[84] By the 1840s, Boston had long outgrown these water sources. In 1846 the city began construction of an aqueduct, a reservoir, and a sixty-mile piping system, all fed by Lake Cochituate (formerly Long Pond) fifteen miles away. The planners promised residents that the system would deliver a reliable, clean water supply to satisfy the city's needs. The plan included street hydrants, public baths, ornamental fountains, and, most important of all, "the protection of your property from the ravages of fire."[85] The aqueduct delivered on its promoter's promise: by the opening ceremony in October 1848, the aqueduct had indeed brought enough water for drinking, bathing, and extinguishing fires. Though Boston's citizens had no view of powerful engines and waterwheels as Philadelphians did, they had ample visual evidence of how technology could be used to harness a natural resource. Reservoirs constructed in several parts of the city, though according to Mayor Quincy "unassuming," were actually massive monumental structures.

Just as New Yorkers had celebrated the Croton Aqueduct six years earlier, the opening day ceremony and celebration in Boston demonstrated how the Boston community not only valued water but also appreciated and celebrated the technology that made the system possible. The civic procession through the streets of Boston to the Common represented the areas of life that the aqueduct improved.[86] As the water commissioner Nathan Hale expressed it, Boston was assembled to celebrate "the accomplishment of a great public work which is of equal benefit to every citizen."[87] Physicians, members of the Overseers of the Poor, merchants, mechanics, and industrial workers, carrying emblems of their craft, marched in the procession. The workers who built the aqueduct walked behind two large water pipes on a platform drawn by seven horses. Another platform held the bricklayers, with the tools of their trade, followed behind by men who had worked on the aqueduct.[88] The city's tailors carried a banner showing Adam and Eve "as a specimen of the human race before the invention of their craft." A group of young women "represented the manufactory of artificial flowers."[89] Practically every civil servant and civic organization in the Boston area was present, including authorities from the towns through which the aqueduct passed, legislators, lawyers, officers and governors of Harvard College, and the Society of Cincinnati. Printers distributed copies of a song composed for the event, "A Song for the Merry-Making on Water Day," struck off from a press carried in a wagon.[90] Temperance societies were well represented. They had begun their celebration of the water as the beverage of choice two years earlier when the city voted to fund the aqueduct; on July 4, 1846, the Boston Total Abstinence Society celebrated the victory over alcohol "by hauling three thousand gallons of water from Long Pond to Boston Common and distributing 'Nature's beverage' to thirsty bystanders."[91]

Mayor Quincy's speech on Boston Common hailed the skill and technology necessary to design and construct the aqueduct. He paid fulsome tribute to the many mechanics' associations—the Mercantile Library Association, Mechanic Apprentices' Library Association, Worcester County Mechanics' Association, and the Masonic Lodge. These organizations had provided the education and information needed by the engineers and mechanics who designed and built the water system.[92] Quincy encouraged his audience to appreciate the aqueduct's symbolic as well as practical importance to the city. As the mayor described the next phase of construction, a reservoir in South Boston, he linked the project to the American Revolution: "The land thus purchased is upon one of the beautiful heights of Dorchester rendered celebrated by a memorable event of the war of the revolution and connected inseparably with the name of Washington." Quincy remarked that this bond with the nation's founding was "highly appropriate for the immediate objects for which this structure is designed."[93] In building a water system, the city constructed an important monument on the scale of those of Greece and Rome. Former president John Quincy Adams acknowledged this connection in his toast at the aqueduct's groundbreaking: "The waters of Lake Cochituate—May they prove to the citizens of after ages, as inspiring as ever the water of Halicon to the citizens of ancient Greece."[94] Adams said that all credit was due to the "energy, the genius, and the skill, of the Water Commissioners, the Engineers, and the Contractors; and they have erected for themselves, in this work, a monument as permanent as the blessings they have secured."[95] The aqueduct and its reservoirs were destined to be long-lasting expressions of an American civilization: "Like the generations of men, a constant succession in this stream will make it permanent, and we cannot but believe, that they who centuries hence, occupy the three hills of Boston, will look back with gratitude to the men of this age, whose foresight and energy secured an unfailing wellspring, for themselves and their descendants."[96]

Just as William Rush's *Water Nymph* at Centre Square and the Croton Fountain at City Hall Park served as visual reminders that hydroengineering fulfilled urban water needs, on Boston Common the water system fed a large fountain erected in the middle of Frog Pond. After the speeches concluded at the opening ceremony, Mayor Quincy turned on the Cochituate water supply and an eighty-foot stream of water burst into the air. The fountain quickly became the centerpiece of the Common, a recreational space removed from the noise and traffic of the city.[97] The cover illustration of the composer George Schnapp's sheet music, the "Cochituate Grand Quick Step," emphasized the fountain as a symbol of the benefits of the water system: the fountain shoots jets of the water that supply the city, while men, women, and children walk and play nearby. Higher than the fountain, an American flag flies from a pole in the foreground. Schnapp dedicated his tune to the mayor, aldermen, and Common Council, thus emphasizing how urban planning and modern technology benefited Boston's citizens. The flag was a potent reminder that such endeavors were part and parcel of the growing American Empire.

FIGURE 4.7. Geo. Schnapp, "Cochituate Grand Quick Step" (Boston: Stephen W. Marsh, Piano Forte Maker and Music Dealer, 1849). Library of Congress, Music Division.

Americans needed the technologies that allowed them to develop industries, transport American-made products, and protect the health and homes of urban citizens. The most acclaimed public endeavor of the early nineteenth century was the Erie Canal. But the most tangible evidence of public works projects were the water systems undertaken in American cities between the 1810s and 1850s. Born out of necessity, these structures were the result of American technological know-how. The magnitude of these projects made them tourist attractions; travel guides that waxed lyrical about the United States' natural wonders, such as Niagara Falls, were equally euphoric about water systems. Philadelphia's Fairmount Water Works and New York's Croton Aqueduct were on the not-to-be-missed tourist's agenda. Lithographs, paintings, sculpture, and music celebrated the grand civic projects. These urban spaces invited men and women to enjoy leisured moments in the middle of a bustling city, while simultaneously appreciating the sites as the triumph of American ingenuity. Moreover, these brick-and-mortar constructions visually communicated imperial ambitions through their design and embellishment. Reservoirs, pumping stations, and fountains articulated Americans' conception of nationhood, and explicitly linked ancient realms with the new American Empire.

CHAPTER 5

Internal Improvements

Phrenology as a Tool for Reform

At the same moment that city water systems offered a visual vocabulary for the expression of American ambitions and ideals, the science of phrenology acquainted men and women with a new way to understand the human mind. Phrenology arrived in the United States in the early nineteenth century and flourished widely for a few decades. What was the significance of this brief but intense fascination with mental science? Phrenology was a diagnostic tool; practitioners claimed that an individual's character and talents could be determined by examining the size and shape of the head. Hence phrenology was the ideal science for the preoccupations of the antebellum era; riding the crest of reform initiatives that emphasized human perfectibility, phrenology dovetailed nicely with temperance, prison reform, education, and health reform. The principles of phrenology were widely disseminated through publications, itinerant practitioners, and visual exhibitions, enabling this democratic utopianism to wrest authority from professionals: phrenology emphasized a layperson's ability to understand and care for their own physical and mental health. Phrenology also gave Americans a new lexicon with which to discuss important issues of the day; because of its saturation in popular culture, the science could be a weapon for satire, or a tool to validate or to challenge long-standing assumptions, especially about race. Although Native Americans received a certain amount of attention and study from phrenologists, for the majority of Americans, Indigenous people were distant and unseen. If white Americans thought about native peoples at all, it was as a barrier to western settlement. Reports on ancient American artifacts often chose to interpret remains of Indigenous civilization as belonging to an extinct people—not the living owners of western land.[1] African Americans, on the other hand, were highly visible. Whether in town or country, white Americans and Black Americans encountered each other daily. In the antebellum era, this familiarity with men and women of a different race encouraged both proslavery advocates and abolitionists to employ phrenological evidence to serve their cause.

Americans and the Science of the Mind

Public lectures in the 1830s spurred the popularity of phrenology in the United States. Prior to that time, curious American physicians and a handful of laypersons read, discussed, and collected texts and objects related to the new science. Americans' interest was based on firsthand encounters with the European theorists and practitioners and with British publications for sale at bookshops and publishing offices in the United States.[2] This American vanguard knew that phrenology, first promoted by the Viennese physician Franz Joseph Gall in the 1790s, offered a radical way of thinking about the mind and body and the connection between the two. Gall based his theory on the idea that the brain is the organ of the mind, an organ composed of twenty-seven innate faculties. The power of a specific faculty depended on its size. Gall drew on earlier studies that suggested the skull takes its outward shape from the shape and size of the brain. Gall's twenty-seven faculties (later phrenologists added to this number) were categorized into five groups: Affective (such as Amativeness and Destructiveness), Sentiments (Love of Approbation, Veneration), Intellectual (Hearing, Sight), Perspective (Form, Number, Tune), and Reflective (Comparison, Causality). The skull became a map, charting exactly where these faculties lay and how large they were.[3]

Widespread American enthusiasm for phrenology had to wait for European lecturers to arrive in 1830s, but curiosity about the new science had already been aroused by the beginning of the nineteenth century. Americans read extracts and reports in imported general magazines such as *Blackwood's*, they purchased books, and they received accounts from Americans abroad. The Philadelphian Nicholas Biddle (1786–1844) was one such early eyewitness. The eighteen-year-old Biddle traveled to Europe as an unpaid secretary to General John Armstrong, the United States minister to France in the fall of 1804. Biddle took full advantage of his European sojourn. He thought of

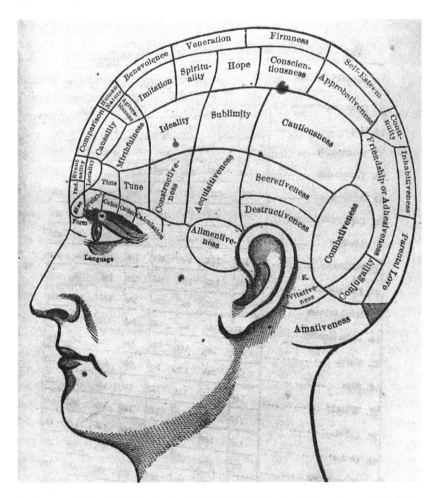

FIGURE 5.1. O. S. and L. N. Fowler, *New Illustrated Self-Instructor in Phrenology and Physiology* (New York: Fowler and Wells, 1859). Author's collection.

himself as a "citizen of the world" open to new ideas such as phrenology. In Germany, he had the opportunity to speak with Gall, whom Biddle confided to his diary "has nothing of the Quack about him. . . . When my head is occupied about any single object I am wholly absorbed by it and attend to nothing else. Thus at present I think only of Dr. Gall and his new system."[4] Biddle took away more than just a remembered conversation with Gall. He returned to the United States in 1807 with a human cranium marked with the location of phrenological organs.[5]

The general public's enthusiasm, or at least curiosity, about phrenology greatly depended on exposure to the science through lectures and demonstrations at which phrenologists provided convincing readings of personal character. Most people accepted phrenology as a legitimate science because it confirmed what they already believed about themselves or their friends and neighbors. The transition from mild curiosity to growing interest and acceptance of phrenology is demonstrated in the diary entries of Christopher Columbus Baldwin, librarian for the American Antiquarian Society in Worcester, Massachusetts, from 1827 to 1835. Between 1829 and 1835, Baldwin's diary records his first encounters with the science and his increasing involvement with Boston area phrenology enthusiasts, culminating with his participation in founding the Worcester Phrenological Society in 1834. In late March 1829, Baldwin attended a whist party at which the guests spent the evening playing cards but also "feeling each others heads: find some well developed *Bumps*."[6] Baldwin acknowledged that he was "disposed to embrace" phrenology "to a limited extent."[7] At the recommendation of his friend Samuel Hoar, Baldwin read George Combe's recently published *The Constitution of Man Considered in Relation to External Objects*. He commented: "Like it much."[8] In addition to reading, Baldwin conversed with several men who embraced phrenology's principles, including the physician Amariah Brigham, the educator and phrenology lecturer William Bentley Fowle, and the painter Chester Harding. Harding told Baldwin that he had taken measurements of "all the most distinguished heads in the country," including Daniel Webster and Chief Justice John Marshall.[9]

Baldwin's interest in phrenology intensified when he made the acquaintance of Harvard's professor of elocution, Jonathan Barber. Barber, according to Baldwin, was also "eminently distinguished as a Lecturer on Phrenology." Baldwin accompanied Barber to the local hospital to observe him treating the "bumps" of patients. Barber also examined Baldwin's head: "He felt of my head and quite surprised me with the information that I had the true developments of the Antiquarian taste, which consisted in the organs of veneration, benevolence and acquisitiveness."[10] Baldwin attended Barber's phrenology lectures a month later, noting that Barber had "about two hundred and fifty to hear him at one Dollar each." Baldwin confided to his diary: "I am a convert to his doctrine to a certain extent. The weight of evidence in favor of the correctness of the doctrine is too great to be thrown down by ridicule. It must be put down as it has been put up, by facts and arguments."[11] Baldwin was drawn ever closer to the circle of Worcester gentlemen who shared his interest in phrenology. In May 1834 he attended a meeting at the Central Hotel to plan a phrenology society. Many of Worcester's leading physicians, including Samuel Bayard Woodward, the medical superintendent of the Worcester State Hospital (one of the first public hospitals for the mentally ill), and several prominent businessmen, including Stephen Salisbury and Anthony Chase, were present. Baldwin wrote: "Like all new converts, we are full of fury and enthusiasm, and we may thank ourselves, if we escape being rank Pagans."[12]

Baldwin maintained a sense of humor about phrenology. In his capacity as Librarian of the Antiquarian Society, he paid a visit to collect books Thomas Walcutt had donated to the AAS. "Seeing that there were something like fifteen hundred volumes before me, my phrenological development of acquisitiveness, not yet satisfied with two tons and a quarter of books and pamphlets, began to enlarge itself and sigh for further accumulations. I could not, however, take advantage of his generosity."[13] Yet, less than a year later, Baldwin's interest had faded. When Nahum Capen, whose Boston firm became the leading publisher of phrenology books in the United States, paid Baldwin a visit in 1835, Baldwin clearly no longer shared Capen's enthusiasm. Baldwin said of him, "He evidently had a strong love for this abstruse, and, to me, uninteresting science."[14] Baldwin's acquaintance with phrenology was typical of the experience of many Americans: reading publications, attending lectures, or, for the adventurous, having one's head read by an itinerant phrenologist. Ironically, Baldwin's interest in phrenology waned just as the science began to reach a broad audience through journals, books, and lecturers who preached phrenology's almost limitless potential to help Americans understand themselves and each other.

Phrenology at Work

Americans who traveled abroad, and those who read the first publications available in the United States, formed the earliest audiences for lectures, demonstrations, and cranial readings. Practitioners of the science benefited from cultural and intellectual developments in the United States in the first decades of the nineteenth century: lyceums, museums, and a variety of new societies and mutual improvement associations all provided ready-made audiences for itinerant speakers. Activities and reportage built on each other: newspaper accounts of a previous evening's demonstration by a visiting phrenologist reached readers in other locales who in turn became audiences when the lecturer arrived in their city. The English traveler and phrenology lecturer Robert H. Collyer noted the abundance of these venues in the early 1840s: "The only recreation offered for the good were lectures, lectures, lectures! The whole region of literature from the days of Queen Bess, up to modern Grahamism was explored to give room for swarms of lecturers upon every *ology* and *ism* that could possibly be thought of."[15]

Charles Caldwell may have been the first itinerant phrenologist in the United States. Following the publication of his *Elements of Phrenology*,

Caldwell set off on a tour of the Northeast in 1828.[16] Little is known about the size of Caldwell's audiences, or even, with the exception of Boston, where he lectured.[17] But just four years later, the German physician Johann Spurzheim's lecture tour in America drew enthusiastic crowds of physicians and laypeople. As Spurzheim traveled through towns and cities, curious individuals, like Christopher Columbus Baldwin, eagerly made his acquaintance. The response to Spurzheim's brief lecture tour was testimony to the scientific and popular interest in phrenology.

Spurzheim arrived in New York City from Paris in August 1832. On his journey to Boston, where he was scheduled to give a series of public lectures, Spurzheim stopped in New Haven, where he met with Yale professor Benjamin Silliman and dissected a brain before a group of faculty members.[18] From New Haven he traveled to Hartford, where he accompanied Dr. Amariah Brigham on a visit to the Asylum for Deaf and Dumb and the Retreat for the Insane and State Prison at Wethersfield. Spurzheim took readings of several of the prisoners there. Brigham told a correspondent "The Warden of the Prison has repeatedly assured me that Dr. Spurzheim gave the characters of many of the criminals, especially of the noted ones, as correctly as he himself could have done who had long known them."[19] These American encounters with Spurzheim confirmed, among those already inclined to do so, that phrenology was an accurate and useful tool. This kind of eyewitness testimony helped promote phrenology in the United States.

Spurzheim's first appearance as a public lecturer in the United States was at the American Institute in the Representatives Hall of the State House in Boston.[20] He then gave several lectures to the Harvard medical faculty, which included John Collins Warren. Warren showed Spurzheim "all the attentions due to a scientific stranger. He examined all my crania." Warren also noted rather dismissively that Spurzheim "afterwards gave a course of lectures on phrenology to a promiscuous assembly of ladies and gentlemen."[21] The Boston *Daily Evening Transcript* announced Spurzheim's upcoming lectures along with the information that tickets were available for purchase at Marsh, Capen and Lyon's Bookstore on Washington Street. Fifty cents purchased admission to a single lecture.[22] The prospect of Spurzheim's lectures excited public interest. One eyewitness recalled that "all had a desire to see and hear him, and the occasion brought together a large and most respectable audience of ladies and gentlemen. . . . The audience seemed to be perfectly delighted. His views were original and practical and all could understand them."[23] Ticket purchasers were given a printed syllabus for Spurzheim's Demonstrative Course of Eighteen Lectures on Phrenology. Spurzheim walked his listeners step by step through the purpose and value of phrenology: the science offered a key to understanding, and then improving, individuals. To engage his audience more fully, Spurzheim employed a variety of visual aids—diagrams, skulls, and casts—to help make the abstract concepts concrete. Unlike Spurzheim's lectures to medical professionals, there were no brain dissections at these lectures. The public had no taste for such visceral displays, nor for the reminder that brains came from bodies that may have been acquired illicitly. Instead, charts, casts, and clean white skulls were on view to explain the new science of the mind.[24]

Spurzheim's lectures were a great success. The Boston publisher of Spurzheim's works and many other phrenological titles, Nahum Capon, recalled that Bostonians were eager to hear Spurzheim. His lectures "attracted alike the fashionable and the learned, the gay and the grave, the aged and the young, the skeptic and the Christian. . . . Some of those who at first attended with a view to collect materials for amusement, or for ridicule, were among the earliest to become converts to his system."[25] The Boston lectures were so successful that Spurzheim began a second series in Cambridge even before the Boston series concluded. John James Audubon, in Boston at the time, hoped to entice Spurzheim to extend his lecture tour to Philadelphia. Audubon wrote, "But the good man is doing so well here, that I suspect he will remain as long as the money pours into his pocket in a way it does at present. His lectures are well attended, and very interesting. He is about to commence a second series on the conclusion of which he will depart for Salem, &c."[26]

But Americans' contact with Spurzheim was brief: three months after he arrived in the United States, Spurzheim died of typhoid fever. Bostonians gave him an elaborate funeral, with nearly three thousand mourners in attendance. Spurzheim was the first person buried in the new Mount Auburn cemetery, whose entryway was an Egyptian Revival temple.[27] In the true spirit of phrenological inquiry, John Collins Warren, attended by other Boston physicians, performed a postmortem examination on Spurzheim at the Harvard Medical College. Spurzheim's skull and brain were removed and preserved for study—at Spurzheim's request—so his phrenological character could be accurately identified.[28] Spurzheim's brain and heart were preserved at the newly formed Boston Phrenological Society.[29] As planned on the day of Spurzheim's funeral, the Boston Phrenological Society acquired a substantial object collection, primarily purchased at the sale of Spurzheim's effects. Various members of the society, including Jonathan Barber, put this rich trove of skulls, casts, and diagrams to use as they lectured to audiences in New England and beyond.[30]

If Spurzheim's American visit ignited popular interest in phrenology in 1832, George Combe's visit six years later confirmed phrenology's popularity in the United States. Though a lawyer rather than a physician, Combe published and lectured on the principles of phrenology. His most influential work was the *Constitution of Man Considered in Relation to External Objects* (1828). Though not strictly a work of phrenology, the *Constitution* presented a theory of natural philosophy which posited that man, just like the rest of nature, was subject to natural laws. Any reform of society or of individuals—be it governance, health, or work—first required an understanding of human nature and these natural laws. Phrenology was the key that unlocked nature's secrets. The *Constitution of Man* laid out a science of man: the phrenological organs were man's constitution. Phrenological knowledge enabled individuals to assess, learn, and educate themselves and others about appetites, passions, instincts, and propensities. Combe placed a powerful tool in the hands of his readers—one that could guide them to understand themselves, and to reform others.[31]

By the time Combe arrived in New York in the fall of 1838, Americans had already become familiar with his writings. Newspapers advertised his phrenology books. Societies and lyceums requested that Combe lecture to them. His itinerary included the principal towns and cities in the Northeast where he delivered a series of public lectures and held private meetings with various individuals—American phrenologists, physicians, educators, artists, and politicians. As Spurzheim had done, Combe visited schools, prisons, and insane asylums. The Boston Phrenology Society honored Combe by inviting him to examine the skull and preserved brain of Spurzheim, the man who had inspired him.

Combe lectured in the principal venues of Boston (the Masonic Temple), Philadelphia (the Museum and then Masonic Hall), and New York (Clinton Hall). The learned and the curious showed up to hear what he had to say.[32] One exception was John Collins Warren in Boston: Dr. Warren chose not to attend Combe's lectures because they were aimed at laypeople, not physicians. Warren thought the lectures had too much entertainment and not enough science: "I never attended his lectures; for I found that, in all the phrenological courses which I attended, the principal object of phrenological lecturers was, not to expose the ground and basis of phrenology, but to interweave it with popular and interesting topics. However judicious this might be, it was, of course, not calculated to give me the information I desired."[33] Warren's opinion was clearly not shared by others. Plenty of people wanted to be entertained as well as enlightened, Nicholas Biddle among them. Biddle's early interest in phrenology persisted into late middle age: when Combe arrived in Philadelphia in the spring of 1839, Biddle presented him with the marked skull he had kept for over thirty years.[34] Combe was pleased with his reception in the United States, and he was optimistic that phrenology would flourish there: "I proceeded thither with the impression that this science would contribute powerfully to the advancement of civilization in that country; and I returned, not only with the impression converted into conviction, but further persuaded, that, in the United States, probably earlier than in any other country, will Phrenology be applied to practical and important purposes."[35] The ever-increasing numbers of practicing phrenologists confirmed Combe's prediction.

Within five years of Combe's visit, the name most closely associated with phrenology in the United States was Fowler. Orson and Lorenzo Fowler became the nineteenth century's leading phrenology lecturers, authors, and publishers. Orson credited his classmate at the Amherst Academy, Henry Ward Beecher, with introducing him to the science: "In 1833, I borrowed Combe's Elements of Phrenology," and a phrenological bust, from my classmate, Henry Ward Beecher, and began its study in right down good earnest, without a teacher, but with zeal."[36] Orson became an itinerant lecturer soon after graduation in 1834. Living in Brattleboro, Vermont, Orson witnessed a former classmate deliver an unsuccessful series of lectures on the battles of the American Revolution. Though Fowler did not say so, he must have been aware of the market for itinerant speakers—the readiness with which communities attended lectures on almost anything. Seizing this entrepreneurial opportunity, he wrote "[I] bought paper, hired a printer, and got out a thousand copies, along with my handbill; ordered a bust, and thirty-two dollars' worth of works on Phrenology, opened my lectures, threw out my card, charged men twelve and a half cents for a phrenological chart, marked, and ladies and children six and a quarter cents."[37]

Orson Fowler knew people liked to hear about themselves—especially if what they heard was flattering. Such satisfying assessments of personal attributes helped spread Fowler's reputation: a reading of the newspaper editor Samuel Smith during Fowler's lectures at Clinton Hall in New York City in December 1836 earned the phrenologist ample praise in Smith's anti-Catholic newspaper, the *Downfall of Babylon, or, The Triumph of Truth over Popery*. Smith reported: "He [Fowler] observed that the phrenological structure of my head was such that it would lead me to take a decided and prominent stand in society. 'I suppose,' said he, 'that you are combating in that you regard the cause of truth against error. You take a bold stand; you fear nothing: you rather court opposition than shrink from it. You are given to deep investigations; persevere through every obstacle."[38] Smith noted that "many are they who enter the Hall entirely incredulous in the science, who return fully convinced of its truth; ourselves were of this number."[39] Smith challenged his readers: "To any one who is skeptical in regard to the science,—we would say, 'Come and see.'"[40]

Scores of people did just that. Newspapers from Maine to Texas advertised books, busts, charts, and the services of phrenologists. A Portland, Maine, bookshop offered Calvert's *Illustrations of Phrenology* with woodcuts.[41] In 1842 a Doctor Hernis (from Paris) set up his office at Mrs. Middleton's boardinghouse in Washington, DC. Hernis informed his customers: "Children's heads examined with regard to their capacities, education, and qualifications for business or professions, at half price." For their convenience (and privacy) ladies could arrange for Dr. Hernis to visit them at home.[42] For aspiring phrenologists, the Keene, New Hampshire bookstore advertised phrenology busts "approved by Dr. Spurzheim" for sale.[43] J. S. Lawrence offered Pittsfield, Massachusetts, customers an exam and a marked chart for 37 ½ cents. A more detailed written exam could be had for 50 cents to one dollar.[44] In Texas, a Professor Smith, at the "urgent solicitation of some of the prominent citizens," of Austin, agreed to deliver a course of phrenology lectures as well as private instruction.[45]

The itinerant phrenologist R. C. Rutherford arrived in Milan, Ohio, in the early spring of 1849. He gave a free lecture at the Hall of the Sons of Temperance that impressed both the town's newspaper editor and the local clergy. With testimonials printed in the *Milan Tribune*, Rutherford began a series of lectures to paying customers.[46] He also made the acquaintance of the wealthy and single Martha Butman. Over the next two years, Rutherford traveled in Kentucky, Indiana, Ohio, Michigan, and Pennsylvania, occasionally returning to Milan where he became engaged to Butman. At some point, Butman began to have doubts about Rutherford both as a potential spouse and as a phrenologist. On March 4, 1851, Butman jilted him at the altar, declaring that Rutherford was "a perjured man."[47] This seems not to have been a case of last-minute doubts. Butman may well have intended this very public scene as an opportunity to expose Rutherford as a fraud: she organized a committee to review the evidence against him. The *Milan Tribune* buzzed with comments and criticisms (under pseudonyms) for the next month. Butman's friends, acting on her behalf, published incriminating excerpts from Rutherford's correspondence—letters that clearly indicated a hypocritical attitude. Even the *Cleveland Plain Dealer* reported the "Row in Church."

Rutherford's public humiliation and exposure was singular. Most locales were eager to avail themselves of the public entertainment phrenologists offered. But phrenology did have its detractors. Ralph Waldo Emerson complained that "Gall and Spurzheim's Phrenology laid a rough hand on the mysteries of animal and spiritual nature, dragging down every sacred secret to a street show."[48] David Reese Meredith of New York put it more baldly: "They have swallowed Maria Monk, abolitionism, and homoeopathia; and are now equally busy in bolting down Phrenology and Animal Magnetism."[49] Certainly the visual displays and mental legerdemain attracted audiences. Many itinerant phrenologists traveled with trunks filled with wall-size illustrations, plaster casts, and skulls. Professor Palmer's lectures in Milwaukee were illustrated "by upwards of 100 diagrams, and busts of distinguished characters, besides a variety of natural skulls, both Indian and European."[50] Phrenologists bombarded their audiences with a multisensory experience: pictures, three-dimensional heads, and demonstrations of phrenology's accuracy. This was phrenology's appeal.

As an eye-catching advertisement explicitly stated: "Call and 'know thyself,' by an application of PRACTICAL PHRENOLOGY to your own head. Get a 'Chart,' and learn to study and modify your evil desires, instead of plunging into every kind of dissipation that would inflame and enrage your 'Animal Propensities.'"[51] Such an alarming threat may have been tongue in cheek, yet many women and men were curious enough to purchase a book or pay for a reading. The earliest phrenology readings were handwritten documents. By the mid-1830s, itinerant phrenologists were traveling with printed forms (some as large as 11 by 20 inches) with a blank for the subject's name and another for the phrenologist's. The simplest charts listed the dozens of organs, organized into categories: propensities, sentiments, perceptive faculties, and reflective faculties, and scored from 1 to 6, 1 to 10, or 1 to 20. The phrenologist wrote the number assigned to each organ to the left of the description. More elaborate charts illustrated exemplary heads of the famous and infamous. Socrates and Shakespeare kept company with more contemporary men such as Franklin, Washington, and Martin Van Buren. Observers could compare their cranial development with these admirable men. The head of "Philip, a notorious thief and liar, in Greenwich st., N.Y." and that of a nameless "idiot," might have given viewers pause for thought.[52] Some charts were embedded in multipaged pamphlets. Orson and Lorenzo Fowler printed their own booklet, *Synopsis of Phrenology; and the Phrenological Developments: Together with the Character and Talents, of* [blank] *as Given by* [blank]: *With References to Those Pages of "Phrenology Proved."* Some one-page charts squeezed a pamphlet's worth of information (in very small print) onto one large page. The phrenologist J. D. L. Zender's "Phrenological chart, or else: a physiognomico-craniological delineation of the person of M _____" provided an explanation of phrenology, included terms used in physiognomy, and illustrated the location of phrenological organs and cranial angles—all on a page twenty-six by nineteen inches.[53]

The phrenology charts and booklets that survive in the archives attest that women and men were interested, or at least curious about, the science. It may not be a coincidence that almost all the readings are positive. Mark Twain recalled that itinerant phrenologists were "one of the most frequent arrivals in our village of Hannibal" in the 1840s, and the citizens who paid twenty-five cents for a reading "were almost always satisfied with these translations of their characters." Twain surmised that this client satisfaction derived from the fact that the phrenologist "was always wise enough to furnish his clients character-charts that would compare favorably with George Washington's."[54] Phrenologists told people what they wanted to hear. They gave readings that tallied with known traits and behaviors. Lorenzo Fowler noted that Charles Dickens, already a famous author when he visited the United States in 1842, "perceives, as if by intuition, the character & motives of men from their physiognomy, conversation & [?]; is suspicious & seldom deceived. Naturally understands human nature."[55]

By design or not, many Americans kept their phrenology readings. Some shared the readings with friends or family. Lucretia Mott wrote to her friend Phebe Post Willis of the readings Orson Fowler did for several Mott family members in 1838: "I don't like the flattery they intersperse in the characters—but tis astonishing how well they describe some sister Eliza's was very well done—so was James's but his was not written out." Mott gave details of her reading: "Nervous temperament—powers of observation not great—rather censorious—arising from great conscientiousness combined with some other organs—strong adhesiveness & moral principles— know your own faults & notice those of others—cannot keep quiet and see things go wrong. As a general thing on other subjects you keep your feelings to yourself, but on moral questions & conduct you do not. The highest possible regard for moral character."[56]

Like Henry Ward Beecher a generation earlier, the young James A. Garfield defended phrenology as a science in a debate while he was a student at Western Reserve Eclectic Institute in Ohio. In 1854 the twenty-three-year-old Garfield visited the Fowlers' office in New York City where Lorenzo Fowler gave him a reading. Garfield recorded in his diary: [I] had my head examined by Mr. Lorenzo N. Fowler. In the main he agreed with others. He said I was inclined to be mentally lazy, and had never called out my powers of mind, that they were greater than I supposed. He told me to elevate my standard of aspiration and thought. I had better aim at the Judge's Bench. Said I needed to be more spirited in resenting an insult."[57] Garfield returned to the Fowlers' office three years later for a second reading. Fowler's second assessment might explain why clients kept their readings. This time Fowler told Garfield he possessed a "Great amount of vitality" and a "Remarkable power of accumulating knowledge." Garfield had a "Bent of mind for Science" and a "Wonderful memory." He also had "a good degree of self esteem." Fowler reinforced Garfield's self-confidence with the comment that Garfield had "the powers and qualities to be a good general. Your mental grasp equal to any task. Can accomplish whatever you undertake and determine to do. Set your mark as high as it can be placed and then work up to it." Fowler advised him that "the profession of the Law for you should only be a steppingstone to something else higher." Garfield clearly took this advice to heart. He became a lawyer, a general, a congressman, a senator, and then president of the United States.[58] The explicit flattery of phrenology readings, which in Garfield's case may explain why he visited the Fowlers twice, encouraged men and women to further acquaint themselves with phrenology's applications: phrenology could aid in choosing marriage partners or business partners. It could assist schoolmasters in educating students. But even for the mildly curious, opportunities abounded to become acquainted with phrenology. Public entertainments, fiction, poetry, and song offered praise for, and criticism of, the intriguing science of the mind.

Phrenology at Play

> Phrenology found many a bump on a man's head and it labeled each bump with a formidable and outlandish name of its own. The phrenologist took delight in mouthing these great names; they gurgled from his lips in an easy and unembarrassed stream, and this exhibition of cultivated facility compelled the envy and admiration of everybody. By and by the people became familiar with these strange names and addicted to the use of them and they batted them back and forth in conversation with deep satisfaction—a satisfaction which could hardly have been more contenting if they had known for certain what the words meant.[59]

Mark Twain's pronouncement expressed his opinion of the gullibility of men and women and his disparagement of phrenology. A popular assault on the science began almost as soon as the first lectures and publications. Demonstrations in lecture halls, museums, and private parlors spurred fiction writers, political commentators, playwrights, and cartoonists to satirize, scorn, or deride phrenology. The fact that the science permeated popular culture is testimony to its share of attention in the public consciousness. More than simply a target for jest and criticism, phrenology provided a platform from which social and political commentaries were cloaked—further proof of the saturation phrenology achieved in American culture.

The prevalence of phrenology in social and cultural venues inevitably prompted American authors to respond. Some writers were enthusiastic promoters of the science. Others were, at best, skeptics and, at worst, severe critics. The attitude writers adopted toward phrenology often depended on their personal experiences with the science. Nathaniel Hawthorne first encountered phrenology while a student at Bowdoin College in the 1820s. Several of Bowdoin's medical college instructors, including Dr. John Doane Wells, whose lectures Hawthorne attended, were early and eager disciples of Gall. Though Hawthorne did not use phrenology in his fiction, he was well enough acquainted with the science to present it to readers of the *American Magazine of Useful and Entertaining Knowledge* when he became its editor.[60] In the April 1836 issue, Hawthorne printed an illustrated phrenology head and an explanation of the organs. Curiosity about phrenology inspired many journals to print essays and illustrations about the science. Hawthorne may well have sensed that the *American Magazine* should not lag behind other popular periodicals in presenting the latest news and information. But Hawthorne's comments at the beginning of the article reveal his skepticism: "In the present state of the science, we cannot advise our readers to expend any considerable part of their time in the study of it. . . . Phrenology . . . must still be ranked among the doubtful sciences."[61] His description of the functions of the phrenological organs confirms his lack of seriousness.

Hawthorne provided his own humorous take on the behavior associated with various organs. Destructiveness: "When strongly developed, it impels people to pinch, bite, scratch, break, tear, cut, stab, strangle, demolish, devastate, burn, drown, kill, poison, murder, and assassinate." Combativeness was the organ "of quarreling and fighting—propensities which seem more likely to produce bumps on the head, than to be caused by them." Mirthfulness: "Before a person tries to be witty, he should examine his head in search of this organ. If he do not find a prominence on the side of his forehead, towards the upper part, he may relinquish all hopes of exciting a laugh, save at his own expense." With tongue firmly inserted in cheek, Hawthorne described Eventuality: "This organ forms a prominence in the middle of the forehead, when largely developed. Individuals who possess it are attentive to all that happens around them. It is essential to secretaries, historians, teachers, and editors."[62]

Herman Melville also employed phrenology for comic effect. In *The Confidence Man* (1857), a potential employer consults the "Philosophical Intelligence Office" where phrenology charts confirm a young man's vocational skills:

> "As for the boy, by a lucky chance, I have a very promising little fellow now in my eye—a very likely little fellow, indeed."
> "Honest?"
> "As the day is long. Might trust him with untold millions. Such, at least, were the marginal observations on the phrenological chart of his head, submitted to the prospective employer by the mother."[63]

Melville's fictional encounter mirrored the phrenology publications that promoted self-knowledge as well as knowledge of others. Melville's fictional employer had difficulty finding an honest, hardworking boy. He, like others, believed that phrenological readings could be relied on to accurately assess honesty, talents, and inclinations. Melville's use of phrenology in his fiction reached its climax in *Moby-Dick*, where the narrator, Ishmael, attempts to apply phrenology to the whale: "If the Sperm Whale be physiognomically a Sphinx, to the phrenologist his brain seems that geometrical circle which it is impossible to square."[64]

Mark Twain also understood the comic potential in the public's attraction to phenology. In a short literary sketch, Twain described the character Jul'us Caesar as "a phrenological curiosity: his head was one vast lump of Approbativeness; and though he was as ignorant and as void of intellect as a Hottentot, yet the great leveler and equalizer, Self-Conceit made him believe himself fully as talented, learned and handsome as it is possible for a human being to be."[65] Twain cynically used phrenology as an example of Tom Sawyer's aunt's promiscuous attraction to the fads of the era: "She was a subscriber for all the 'Health' periodicals and phrenological frauds; and the solemn ignorance they were inflated with was breath to her nostrils. . . . She was as simple-hearted and honest as the day was long, and so she was an easy victim."[66]

In contrast to Melville and Twain, Edgar Allan Poe—who, as his Balloon Hoax demonstrates, was quick to exploit a fad—approached phrenology with more respect. Poe first declared his regard for the science in a review of Mrs. L. Miles's book *Phrenology, and the Moral Influence of Phrenology* in 1836. Poe opened his review by asserting: "Phrenology is no longer to be laughed at. It *is* no longer laughed at by men of common understanding. It has assumed the majesty of a science; and, as a science, ranks among the most important which can engage the attention of thinking beings."[67] Despite his tone, Poe may have had a quiet joke with those who knew him personally. His description of a well-developed skull exactly fits his own: "To this may be added the opinion of Gall, that a skull which is large, which is elevated or high above the ears, and in which the head is well developed and thrown forward, so as to be nearly perpendicular with its base, may be presumed to lodge a brain of greater power (whatever may be its propensities) than a skull deficient in such proportion."[68]

Poe, though playful in his review, was serious in his application of phrenology to his own work. He credited the crime-solving abilities of his famous detective, C. Auguste Dupin, to the phrenological organ of Analysis. Poe's most famous story, "The Murders in the Rue Morgue," is macabre, mysterious, and violent—all the elements required in sensational fiction. Poe took pains to guide the reader toward the solution to the baffling mystery through reason and analysis, abilities that he frames within the context of phrenology. The story opens with a long preamble that reads very much like a phrenology text: "It is not improbable that a few farther steps in phrenological science will lead to a belief in the existence, if not to the actual discovery and location of an organ of *analysis*."[69] The narrator's point is to introduce Dupin's mental skills in the context of phrenological theory. Of Dupin, the narrator writes: "At such times I could not help remarking and admiring (although from his rich ideality I had been prepared to expect) a peculiar analytic ability in Dupin. He seemed, too, to take an eager delight in its exercise, if not exactly in its display."[70]

The flirtation that several now-iconic authors of the nineteenth century had with phrenology illustrates how the science grasped the public's attention. As early as 1826, James Kirke Paulding's social commentary, *The Merry Tales of the Three Wise Men of Gotham*, satirized phrenology and phrenologists. Along with the misdeeds of lawyers and Robert Owen's utopian experiments, Paulding created Le Peigne, a convert to phrenology: "It appeared to me impossible, indeed, that a rational being could shut up his understanding to the conviction of its irresistible demonstrations."[71] Le Peigne relates to his listener, Mr. Quominus, how he found both his business partner and his clerk with the aid of phrenology. Le Peigne then looked for a wife: "As this was the most important matter of all, I resolved to be very particular, and to apply the rules of my art with more than ordinary circumspection. In the first place, it was indispensable that she should have a perfect development of the organ of amativeness."[72] To his dismay, his wife "did frequent violence to the organs of order and adhesiveness—for she left my house at sixes and sevens, and seemed to adhere to nothing but her own will." Moreover, Le Peigne's business partner embezzled all the profits, and his clerk ran away.

Le Peigne's wife, who demonstrated amativeness more to her dog than her husband, left him. After Le Peigne relates these dismal events, Quominus responds, "'I suppose this put an end to all doubts as to the infallible auguries of the cerebral development?' 'It did,' replied the other—'it established their truth in my mind beyond all contradiction or question.'"[73] Like Voltaire's Dr. Pangloss, no adverse circumstance ruffles Le Peigne's sublime conviction about the correctness of his theories.

Irrational adherence to theories regardless of overwhelming evidence to the contrary was a mainstay of the comic portrayals of phrenology and phrenologists. Popular theatrical entertainments from the 1820s through the 1840s depicted misguided phrenology enthusiasts. The earliest of these entertainments was *The Phrenologist, or The Organs of the Brain*, performed at Philadelphia's New Theater in 1823. The play ran for only two nights, as a farce accompanying the more somber *Richard the Third*. Like Paulding's phrenologist, the lead character, Baron Rückenmark (the German word for spinal cord), defies all evidence that conflicts with phrenological readings. He misidentifies honest men as thieves and thieves as honest men. Rückenmark is also convinced that his servant, Peter (the choice part played by a comic actor in the Philadelphia performances), is a genius—despite ample evidence to the contrary.[74] The perennial theatrical device of gendered disguise is employed to poke fun at phrenologists: Rückenmark is certain that one of the characters, a woman disguised as a man, really is a man because of the shape of "his" skull. Rückenmark feels the woman's head and declares "O!O!O!—Ha! ha! ha! They have done you wrong, my worthy Mr. Ellstern, much wrong . . . you are no woman."[75] The successful lecture tours of Spurzheim, Combe, and many other itinerant phrenologists in the 1830s and 1840s provided more fodder for comic stage performances. A Philadelphia teenager, J. Warner Erwin, recorded attending Dr. Valentine's performances twice in the space of three days in March 1841. Among Valentine's repertoire was his comic "Yankee lecture on Phrenology."[76] Another troupe, the Croton River Minstrels, included "a burlesque lecture on Phrenology" at their "Grand Concert" in Sing Sing prison in 1848.[77]

Newspapers, magazines, and even seasonal gift books inserted humorous phrenology stories to entertain readers. "The Young Phrenologist," by John Neal, first appeared in the *New England Galaxy and United States Literary Advertiser* in 1835. It must have been a popular piece, because publisher Samuel Goodrich reprinted it in his gift book, *The Token*, for the 1836 Christmas season. Gift books were designed to entertain readers with poetry, songs, and stories that evoked pleasure and laughter. The inclusion of a humorous phrenology tale demonstrates readers' familiarity with the science—even if phrenology was the butt of the joke. The story opens with a young wife discovering miniature phrenology heads in a drawer of her husband's desk, "the five and forty little monsters all staring at her, as if they enjoyed her perplexity." Among these frightening curiosities, she recognizes her own image: "a miniature of herself, with the hair wiped off, and the bare ivory skull, written all over with unutterably strange characters." When she confronts her husband, he responds by taking out the phrenological reading he has done of her and replies "Do you know, my dear, that you are indebted to Phrenology for a husband?" As he lists her traits, Marvellousness is large. She says "No, no, stop there if you please; I don't see how that can be; I don't believe in ghosts." Her spouse responds, "No, but you do in the *Christian Examiner*." He tells her with satisfaction that her Philoprogenitiveness is very large—a sign that she will make a good mother. Annoyed at both his smug confidence in his judgments as well as the realization that science, rather than affection, guided his choice, the wife turns the tables on him and delivers her own summation of her husband's character traits: "Self-complacency—very large—prodigious!" "Audacity—unparalleled!" "Ambition—frightful, inordinate, insupportable." "Modesty—wanting."[78]

One of the most extended phrenology jokes appeared in *Harper's New Monthly Magazine* in 1856. The anonymously authored "January First, A.D. 3000" takes an unnamed Rip van Winkle–like narrator on a tour of Peerless

FIGURE 5.2. "Five and forty little monsters." Bally's Miniature Phrenological Specimens. Science Museum / Science and Society Picture Library.

City in the year 3000. Among other marvels of technology, including flying machines, the citizens of this enlightened republic have perfected child-drearing. Babies are collectively reared in a factory-like system and fed "supra-incto-gone." When they are fifteen months old, the infants must pass an examination before the State Phrenological Commission. As the narrator's guide explains, "Their heads are thoroughly examined, their mental capacities recorded, and their vocation in life decided. On leaving the Commissioners' room, each infant has a ticket pasted on its person, bearing the name of the trade or profession to which it is destined."[79] Like Twain's Philosophical Intelligence Office, the State Phrenological Commission channels individuals toward work for which they are deemed phrenologically suited.

Readers understood the joke because phrenologists focused so much attention on education. They not only promoted phrenology's utility, they emphasized the value of phrenology as a vocational tool. Phrenologists from Spurzheim onward routinely visited schools, measured pupils' heads, and pronounced judgment on particular traits and propensities. Anecdotes of phrenologists' ability to judge a student's talents are sprinkled throughout American phrenology texts. One of the earliest was Spurzheim's visit to William Fowle's Monitorial School in Boston in 1832. As Fowle recalled, "It was astonishing to see with what facility he could point out among the scholars of a school, those who were remarkable for any superiority or deficiency. His quick and penetrating eye seemed to read the very thoughts and feelings of those around him, and his remarks which immediately followed, showed his entire confidence in the truth of his science and the certainty of his decisions."[80] Phrenology as a diagnostic tool for educators was promoted in a variety of venues (most obviously with the school visits), converting teachers, such as Fowle, one at a time. When Horace Mann was appointed to head the Massachusetts State Board of Education, he made George Combe's *Constitution of Man* required reading for future teachers in the state's Normal Schools. Mann considered the *Constitution* to be "the only practical basis for education."[81] As an essay in the *Knickerbocker Magazine* stated the case, "This science enables the teacher to understand the mental capacities of his pupils, and to adapt their studies accordingly. It should decide one in the choice of his profession, and settle upon his walk in life."[82] Phrenology's occupational utility was recognized well into the second half of the nineteenth century: Nelson Sizer's 1872 handbook *What to Do, and Why* linked individuals' talents and traits with occupations that suited them. Sizer listed manufacturing, trades, agriculture, and businesses of various kinds.[83]

Educators embraced phrenology as a predictor of talent, and as an aid to pedagogy. Political commentators, on the other hand, used it as a tool for satire. Employing phrenology's terms and definitions, they offered critical assessments of leading politicians. As early as 1824, almost a decade before phrenology reached a popular audience, an anonymous letter in a Charleston, South Carolina, newspaper referenced phrenology in order to convey admiration for Secretary of War John C. Calhoun: "The whole contour of his head is one of the finest I have ever seen, and a skillful phrenologist would at once pronounce him to be a man of genius."[84] The 1840 poem "To the Great Expunger!" by the pseudonymous "Pindar" praised Senator Thomas Hart Benton of Missouri. Pindar laments that Benton's mother might have predicted her son's future greatness if only she had understood phrenology:

In raptures she'd have hug'd her darling boy;
No more for native *tenants* search'd thy crown,
When "bumps" *momentus*, might her cares employ
"Dear child, these bumps do certify to me,
Thou wilt one day, a 'great expunger' be!"[85]

Politicians themselves coated the barbs thrust at their opponents in phrenological terms. When Henry Clay protested President Jackson's interference with the Senate in 1834, Clay asserted that Jackson's misguided actions were due to his highly developed organ of Destructiveness: "Except an enormous fabric of Executive power for himself, the President has built up nothing, constructed nothing, and will leave no enduring monument of his Administration. He goes for destruction, universal destruction, and it seems to be his greatest ambition to efface and obliterate every trace of the wisdom of his predecessors."[86]

The cartoonist David Claypoole Johnston devoted his collection of satirical commentary, *Scraps for 1837*, to a phrenological analysis of President Jackson's wielding of presidential powers for his pet causes. Johnston's annual publication on American manners, habits, and political practices each sold more than three thousand copies, from Maine to South Carolina.[87] "Veneration" depicts Jackson as the masthead of a ship named Constitution, with men kneeling around him. One man says "Perish commerce! Perish credit perish our institutions! Perish the universe itself if in thy wisdom thou wouldst have it so." Another kneeler says "Oh great preserver of our constitution." The caption quotes George Combe's explanation of the organ: "Veneration when vigorous & blind produces complete prostration of the will & the intellect to the object to whom it is directed & is the cause of every kind of superstition as worshiping beasts & stocks & stones." "Hope" shows Jackson reaching for a crown, with a man clinging to Jackson's coattails. A paper with the word "appointment" hangs from the man's pocket. Behind him another man clings to the first man's coattails and utters "My friend has got an appointment & has promised me an office so I'll hang on." Out of his pocket dangles a paper with "promise of office" on it. A third man hangs onto the second's coattails and says "I have no promise but I voted for my friend here who is always ready to reward his friends when he gets rewarded himself & I think he's got a considerable smart chance." Behind the presidential chair squats President-elect Martin Van Buren, blowing bubbles. The caption is another Combe quote: "This faculty produces the tendency to believe in the possibility of what the other faculties desire, but without giving conviction of it, which depends on reflection."[88]

In "Constructiveness" Van Buren props up an enormous floor plan for Jackson to admire. At the top it reads "Plan of a splendid edifice to be erected in Washington"; at the center is the "presence Chamber" complete

FIGURE 5.3. "Hope." David Claypoole Johnston, *Phrenology Exemplified and Illustrated, with Upwards of Forty Etchings: Being Scraps No. 7, for the Year 1837* (Boston, 1836). Courtesy of the American Antiquarian Society.

with throne; at the side is the "Kitchen Cabinet's private chamber" and the "Depository of U.S. Funds."

Around the same time that Johnston was sketching illustrations for *Scraps*, an anonymous critic of Martin Van Buren published the phrenological measurements of Washington's leading politicians, including John Quincy Adams, John C. Calhoun, Henry Clay, William Wirt, Martin Van Buren, John Marshall, and Daniel Webster. Van Buren's head did not measure up to those of his political opponents:

> The greatest distance from ear to Individuality, which measurement is thought to show the strength of a portion of the intellectual organs, is that of John Quincy Adams, being 53, while that of Martin Van Buren is the least, being only 47. . . . The heads of all measure large between the organs of destructiveness, but that of Van Buren is the largest, being 6 inches 4 tenths—giving an immense breadth between the ears. It is to be regretted that the measurement from Secretiveness to Secretiveness had not been taken. . . . From ear to comparison, showing the length of fibre in the reflective organs, the distance is greatest in the heads of John Q. Adams and Daniel Webster, being in each 5 inches 6 tenths—while on that of Martin Van Buren it is the least, being only 5 inches and 1 tenth.[89]

The author regretted the absence of measurements for Jackson, for whom it would probably be the case that "the distance from the ear to Firmness considerably greater, than in either of the individuals above named—and the distance from Cautiousness to Cautiousness much less."[90] The Whig candidate Zachary Taylor fared no better at the hands of the cartoonists. Henry R. Robinson took advantage of Orson Fowler's phrenological authority to pronounce Taylor unfit for the presidency. In a lithograph published in 1848, Taylor sits for a reading. The *New York Tribune* editor, Horace Greeley, asks Fowler, "What for a President would he make?" Fowler replies, "He says he is 'Incompetent,' & so say his developments." A shelf behind Fowler displays several busts, including Martin Van Buren and Henry Clay. This juxtaposition of past and present political candidates suggests that Taylor might not measure up to his party's (phrenological) standards.

Harriet Martineau, an Englishwoman and keen observer of American culture, noted phrenology's popularity in the 1830s: "When Spurzheim was in America, the great mass of society became phrenologists in a day, wherever he appeared; and ever since itinerant lecturers have been reproducing the same sensation in a milder way, by retailing Spurzheimism, much deteriorated, in places where the philosopher had not been." Martineau described

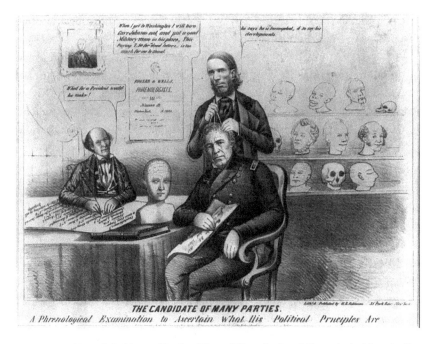

FIGURE 5.4.   Henry R. Robinson, *The Candidate of Many Parties. A Phrenological Examination to Ascertain What His Political Principles Are* (New York: Lithd. and published by H. R. Robinson, 1848). Library of Congress.

enthusiastic Americans pulling off "all caps and wigs," while "all fair tresses [were] disheveled," as men and women examined their phrenological bumps.[91] Practical phrenologists were responsible for acquainting Americans with the science, if only superficially. But this was enough information for people to get the jokes in theatrical entertainments, fiction, and political satires. While it held the public's attention, phrenology permeated American life; education, social reform, family life, and the arts all drew on the science of the mind.

The Science of Race

Phrenology was more than just entertainment. Americans applied phrenology to matters of great concern in the antebellum era—slavery, abolition, and the place of Black Americans in white America. Anglo Americans had long asserted white superiority and Black inferiority, a claim that rested on anecdotal evidence: from travelers' tales that depicted African women as beasts (so much so that they mated with apes) to Jefferson's anthropological observations in *Notes on the State of Virginia*.[92] Americans used phrenology to confirm these long-standing assumptions about race, employing science to validate what most men and women already believed.

Just at the time that Americans were introduced to phrenology in a variety of cultural and entertainment venues, the balance of opinion about racial differences was shifting toward a belief that difference was immutable and unchanging. Hosea Easton and Samuel Stanhope Smith were perhaps the last American defenders of an environmentalist theory of difference in which climate, diet, education, and physical labor all played a role in shaping an individual's character, intellect, and, according to some, skin color. Smith, a Presbyterian minister and president of Princeton University, defended environmentalism because he saw it as a consequence of a unified theory of creation: Christian theology required environmentalism. Smith's *An Essay on the Causes of the Variety of Complexion and Figure in the Human Species* (1810), like many early ethnographies, dealt anecdotally with claims about environmental influences. He drew a sharp contrast between free Blacks in northern states and slaves in the South: "In some of the New England states, for example, we remark, in the body of the people, a certain composed and serious gravity in the expression of the countenance, the result of the sobriety of their domestic education, and of their moral and religious, their industrious and economical habits, which pretty obviously distinguishes them from the natives of most of the states in the southern portion of the Union." To Smith, it was clear that the circumstances of slavery, generation after generation, degraded natural talents and abilities; what scientists saw when they measured, examined, and observed enslaved men and women was the result of centuries of physical and psychological ill-treatment.[93]

Hosea Easton's writing was inspired by a lifetime of personal experience of prejudice against people of color in New England. Easton was a founding member of the National Colored Convention, an organization that aimed to reverse the effects of generations of discrimination through an ambitious program of "uplift"—establishing schools, literary societies, and temperance organizations in the North.[94] Easton's pamphlet, *A Treatise on the Intellectual Character, and the Civil and Political Condition of the Coloured People of the United States and the Prejudice Exercised towards Them* (1837), underscored the NCC's goals by articulating, from historical examples and religious teaching, that African Americans were the intellectual, moral, and spiritual equals of whites. Easton concluded "It is a settled point with the wisest of the age, that no constitutional difference exists in the children of men, which can be said to be established by hereditary laws."[95]

In contrast to Easton's arguments for racial equality, most phrenologists embraced an essentialist view of humanity in which racial differences were immutable and racial hierarchies were clear.[96] The American physician Charles Caldwell vigorously asserted white supremacy and Black inferiority in *Elements of Phrenology* (1827). According to Caldwell, only the Caucasian race exhibited "real human greatness," while "the genuine African figure occupies an intermediate station between the figure of the Caucasian and the Ourangoutang."[97] George Combe's widely available *A System of Phrenology* compared the "Natural Talents and Dispositions of Nations, and the Development of Their Brains" and found "distinct and permanent features of character which strongly indicate natural differences in their mental constitutions" in Europeans, Asians, Africans, and Americans. Though less forcefully worded than Caldwell's assertions, Combe, too, subscribed to racial hierarchies. Of the Africans, Combe stated, "The annals of the races who have inhabited that Continent, with few exceptions, exhibit one unbroken scene of moral and intellectual desolation; and in a quarter of the globe embracing the greatest varieties of soil and climate, no nation is at this day to be found whose institutions indicate even moderate civilization." This lack of development was discernible through both anthropological and phrenological observation:

> One feature is very general in descriptions of the African tribes; they are extremely superstitious. . . . This character corresponds with the development which we observe in the Negro skulls; for they exhibit much Hope, Veneration, and Wonder, with comparatively little reflecting power. Their defective Causality incapacitates them for tracing the relation of cause and effect, and their great Veneration, Hope, and Wonder, render them prone to credulity, and to regard with profound admiration and respect any object which is presented as possessing supernatural power.[98]

Combe claimed that these deficiencies were made up for by traits that made Africans "polite and urbane, and hence [they] make excellent waiters." Large philoprogenitiveness made them "our best nurses, as far as fondness and patience with children are concerned."[99] Combe argued that Caucasians, in contrast, exhibited "superior force of mental character. . . . In short, they indicate a higher natural power of reflection, and a greater natural tendency to justice, benevolence, veneration, and refinement, than the others."[100] Proponents of African colonization enlisted this phrenological evidence of difference to bolster their cause, asserting that "the white is not only endowed with a larger volume, but with a better organization of brain than the Negro, so that the first has not only more power, but that power fitted for a superior intellectual and moral direction." Colonizationists argued that it was their duty to "remove the temptation to the sin of domination over a weaker brother, by restoring him to the condition for which he was created, instead of making vain efforts to do him justice in circumstances where it is morally impossible, and where it is, therefore, an inconsistency to make it a point of religious duty."[101] Abolitionists placed less reliance on phrenologists' claims of inherent racial difference. In a speech delivered before the House of Representatives during the debate on abolition of the slave trade in the District of Columbia in 1837, the Whig congressman William Slade expressed his skepticism: "Differences of intellect. *What* differences? How are they to be *defined*? The science of phrenology may perhaps, by and by, furnish some aid; but in its present imperfect state it can hardly be trusted with so grave a matter as this."[102] Nonetheless, environmental theorists were increasingly drowned out by those who asserted essential, and unalterable, distinctions between the races, and by the results of a decade of violence aimed both at northern Blacks and at whites intent on uplifting free people of color.[103]

Attacks on free people of color in the North began with the advent of gradual emancipation policies in the 1780s. Black communities in the postrevolutionary era increasingly displayed attributes of a stable society: they built churches, schools, and successful businesses. They organized mutual benefit societies. Black laborers, always a visible presence in port cities, increasingly worked side by side with whites.[104] Blacks paraded, sometimes in military-style uniforms, in northern towns and cities.[105] By the first decades of the nineteenth century, discomfort with the presence of free people of color in northern towns and cities had been bolstered by the American Colonization Society's agenda to rid the country of nonwhite Americans; the Black race and the white race "must live forever separately and unequally" on opposite sides of the Atlantic.[106] But as some reformers denounced the ASC's policies and demanded an immediate end to slavery, white objections to the presence of Blacks in northern communities became more frequent and more violent.[107] The abolitionists William Garrison, Arthur Tappen, and Simeon Jocelyn inflamed local opposition to the integration of Blacks and whites in New Haven with their plan, in conjunction with the National Colored Convention, to open a college for Black men in 1831. Word of the proposal provoked the town's white citizens to throw garbage at Jocelyn's house and to attack New Haven's Black neighborhood, known as "new Liberia."[108] Integrated schools in New England were also targeted: Prudence Crandall, threatened by mob violence and by colonizationists' legal harassment, abandoned her plan for a girls' school in Canterbury, Connecticut. The Noyes Academy in Dover, New Hampshire, was destroyed by a mob. Whites attacked Black neighborhoods in Boston, Providence, New York City, Hartford, Pittsburgh, and Utica. In 1836, a year before he published the *Treatise*, Hosea Easton's church in Hartford was burned to the ground.[109]

In the context of this campaign of intimidation, phrenologists joined the national conversation about racial difference with visual displays of the "good Negro": heads and skulls at lectures and museums that emphasized the phrenological claims of the relationship between physical traits and personal behavior. By the early 1840s, the Fowlers were selling sets of casts at their New York office. Twenty-five dollars bought thirty-nine heads, masks, and skulls. The majority of the ceramic busts were of prominent individuals such as Napoleon, Voltaire, Sir Walter Scott, Aaron Burr, and Henry Clay. Casts were either made from live sitters or from sculpture. Each came with phrenological evaluations demonstrating each subject's talents. All the casts that were made directly from skulls, on the other hand, illustrated oddities, deficiencies, or exemplary examples of phrenological composition. Among the latter were Patty Cannon, Murderess ("All the Moral organs small"), a "Carib" ("An untamable savage, and of the lowest order of human beings"), one Tardy, a Pirate ("All the selfish organs, very large"), and a "Good Negro, a slave—Selfish Organs, small. Moral, Social, and Intellectual organs, large."[110] Another "Good Negro" profiled in the Fowlers' journal was Eustace, "eminently distinguished for qualities of virtue and benevolence." A slave in Haiti, Eustache was credited with saving his master and several hundred other white Haitians during massacres carried out under Jean-Jacques Dessalines in 1804. Taken by his owner to France, Eustache taught himself to read so he could read aloud to his elderly and increasingly visually impaired owner. According to the author of an article in the *American Phrenological Journal*, Eustache "secretly applied himself to study; took lessons at four o'clock in the morning, in order that the time necessary for the performance of his regular duties might not be encroached upon; speedily acquired the wished-for knowledge; and, approaching the old man with a book in his hand, proved to him, that if nothing seems easy to ignorance, nothing is impossible to devotion." Although freed after his master died, Eustache "has always preferred to remain in the condition of a servant, in order that he might turn to account his skill in cookery, and enable himself to do good to his fellow-creatures." The author thought it unlikely that "another such instance of pure virtue and disinterested benevolence can be found recorded in the annals of history. It is the more striking, inasmuch as the individual belonged to a race generally regarded as deficient in those qualities." The key to this remarkable display of seemingly un-Negro-like behavior lay in Eustache's skull: "It will be obvious to every phrenologist, . . . that the head of Eustache was of vary considerable size. In this respect as well as in its form, it has quite the appearance of a European head." Eustache's exemplary cranium was combined with his deferential personality: "In his youth, he was noted for avoiding light and vicious conversation, and for embracing every opportunity of listening to intelligent and respectable whites."[111]

These anecdotes of obedience were concurrent with increasing calls for abolition. George Combe argued that the American slave's nature, measured by phrenological standards, defused antiabolitionist fears that liberty led to rebellion. Combe refuted Henry Clay's argument that emancipation would generate violence: Combe suggested that phrenology proved Black Americans, by their nature, were "essentially amiable." Comparing Blacks to Native Americans, Combe wrote, "The one is like the wolf or the fox, the other like the dog. In both the brain is inferior in size, particularly in the moral and intellectual regions, to that of the Anglo-Saxon race, and hence the foundation of the natural superiority of the latter over both." Combe reassured opponents of abolition that the inherent qualities that "render the Negro in slavery a safe companion to the White, will make him harmless when free. If he were by nature proud, irascible, cunning, and vindictive, he would not be a slave; and as he is not so, freedom will not generate these qualities in his mind."

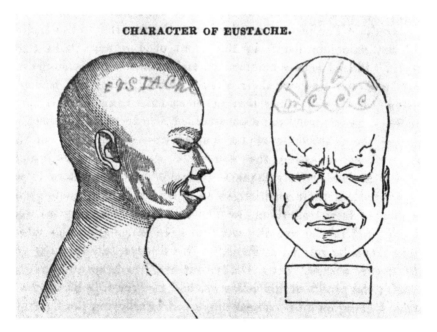

CHARACTER OF EUSTACHE.

FIGURE 5.5. "Article V. Character of Eustache." *American Phrenological Journal and Miscellany* 2, no. 4 (January 1, 1840): 177. Courtesy of the American Antiquarian Society.

Combe was quite explicit: "The fears, therefore, generally entertained of his commencing, if emancipated, a war of extermination, or for supremacy over the Whites, appear to me to be unfounded."[112]

If Jefferson's analogy for slavery was holding a wolf by the ears—the impossibility of controlling a wild creature—Combe's American slave was a domesticated animal who, once free, would be docile and nonthreatening. Combe's description of a hotel worker in Philadelphia confirmed his general assessment of Black Americans: "His manner of thinking, speaking, and acting indicates respectfulness, faithfulness, and reflection."[113] Anecdotal observations of slave behavior reported in the phrenology journals accorded with Combe's opinion: E. B. Olmstead noted that the "part of Veneration exercised in obedience is characteristically large in slaves." A report of a manumitted slave who voluntarily re-enslaved himself noted that this behavior arose from large "Inhabitativeness": "The love of home, or Inhabitativeness, is the true explanation of this negro's conduct. He preferred his home with slavery, rather than freedom abroad."[114] Expressions of wishful thinking about passive and obedient Black Americans were more than simply an expression of southerners' fear of slave violence. They were just as much a response to northerners' anxieties about free people of color.[115]

Captives on Display

Theories about race, and the arguments for and against slavery, were all in play when the US Coast Guard boarded the schooner *Amistad* in August 1839. The abolitionists who formed the Amistad Committee to aid and defend the African captives faced the challenge of gaining the Africans' legal freedom. Hence, the *Amistad* captives' violent act of liberation required a careful public relations campaign: the committee sought to win the hearts and minds of the public. This was no easy task in a society so recently rocked by racial violence and white ambivalence about Blacks living in their midst.

When the US Coast Guard seized the *Amistad* near Long Island, they found fifty-three men and children aboard who had been enslaved in Sierra Leone and transported to Cuba. Two Spaniards had purchased the Africans in Havana and then transferred them to the *Amistad*. On the journey to another destination in the Caribbean, several of the captives escaped their chains, attacked the captain and crew, and ultimately forced the crew members to redirect the ship toward Africa. Instead, the sailors turned the ship northeast. It was in Long Island Sound when the Coast Guard apprehended the *Amistad*. The Coast Guard released the Spanish slave owners and crew and jailed the Africans on murder charges. Although President Martin Van Buren wished to extradite the captives back to Cuba, abolitionists raised money to defend them in an American court. The ensuing trial involved murder charges against the Africans and property claims by the Spanish shipowners. The Federal District Court in Connecticut first heard the case and decided in favor of the *Amistad* captives: they were ruled to be free people unlawfully enslaved and claimed as Spanish property. The US government then appealed the Federal Court decision. When the US Supreme Court heard the appeal in January 1841, it ruled in favor of the Africans. Though eighteen men died before they were freed, eventually the surviving men and children returned to Africa.[116]

The captives spent most of their eighteen-month incarceration in the New Haven jail, a virtual centerpiece for the American abolitionist cause. A steady stream of visitors reported on the character, behavior, and intelligence of the captives. The Africans drew attention and curiosity from the moment they were arrested. Within a few weeks of their arrival in the United States, the Bowery Theater in New York City staged a new play, *The Black Schooner, or, The Pirate Slaver Amistad*. Meanwhile, in New Haven, the Africans were on display both inside and outside the jail. They were allowed exercise on New Haven Green, where "the massive crowds that filed through the jail assembled to watch bodies fly through the air across the green, then followed them back to their cells for another look."[117] When the captives were transported from New Haven to Hartford in September for the first legal hearing, crowds descended on Hartford to view the captives. The antiabolitionist *New York Morning Herald* printed an engraving of the captives in Hartford that showed visiting abolitionists, women and young children, and a phrenologist. Most likely the unnamed phrenologist was George Combe. While in Hartford for a series of lectures in September, Combe visited the *Amistad* captives. His detailed phrenology reading of Cinque—who was thought to have been the key leader among the men who freed themselves and attacked the *Amistad* crew—reveals not only Combe's curiosity about Cinque's personality but his abolitionist sympathies as well:

> Their heads present great varieties of form as well as of size. Several have small heads, even for Africans; some short and broad heads, with high foreheads but with very little longitudinal extent in the anterior lobe. Their leader Cinquez or Jinquez, who killed the captain of the schooner, is a well-made man of 24 or 25 years of age. His head is long from the front to the back, and rises high above the ear, particularly in the regions of Self-Esteem, and Firmness. . . . This size and form of brain indicate considerable mental power, decision, self-reliance, prompt perception, and readiness of action. . . . It is impossible to look without horror and indignation on these young and unoffending men

and children deprived of their liberty, reduced to slavery, and converted into mere 'property,' by *Christians*; I say by Christians, because I have no doubt that if any one were to deny that their reputed owner, who also is here, or his advocates in the American press, were Christians, he would be prosecuted for a libel on their religious character![118]

Despite his strong feelings about slavery, Combe did not openly speak in favor of abolition during his American tour. He did not make his comments on the *Amistad* captives public until his travel memoir was published several months after the captives' release. But Combe did take pains to assure his readers (after the fact) that, "as the subject lay incidentally in my way, I have not shrunk from it, but have introduced the skulls and casts of Negroes among those of other varieties of mankind, and freely expressed my opinion of the moral and intellectual capabilities indicated by their forms." Combe went so far as to describe slavery as "a cancer in the moral constitution" of Americans.[119] In a private letter to the abolitionist Dr. W. B. Sprague of Albany, New York, Combe declared that slavery was the "darkest stain on the fair face of American freedom."[120]

Lorenzo Fowler's reading of Cinque, on the other hand, was intended for immediate publication. Around the same time the Bowery Theater opened *The Black Schooner*, Fowler rushed to the New Haven jail. He published his report on Cinque in the next issue of the *American Phrenological Journal*. Fowler found Cinque's intellect "better developed than most persons' belonging to his race." Moreover, Fowler compared Cinque favorably to African Americans: "His cerebral organization, as a whole, I should think, was also superior to the majority of negroes' in our own country." Unsurprisingly, Fowler's analysis of Cinque coincided with Cinque's role in the *Amistad* rebellion. Fowler's report was much more detailed than Combe's and more closely tied to the events on the *Amistad*. Fowler concluded that Cinque "would have great self possession in times of danger, and might easily conceal, by the expressions of his countenance, all appearance of his real feelings or designs, so that it would be difficult to find him out, or detect his plans." Cinque's phrenological faculties "admirably adapt him to take the lead, secure power, and command the respect of others as well as render him capable of exerting a controlling influence over the minds of those like the native Africans." Fowler discovered that Cinque's faculties gave him "a love of liberty, independence, determination, ambition, regard for his country, and for what he thinks is sacred and right . . . joined with at uncommon degree of moral courage and pride of character." Furthermore, though Cinque was not "revengeful or ill-natured, he ha[d] too much pride and love of self to become subject to the will of others."[121]

Fowler took a plaster cast of Cinque's head during his visit and used an illustration of the cast in his *American Phrenological Journal* article. Readers were told that the cast was on display, and copies of it were for sale, at the Fowlers' New York office. By the following November, at least one itinerant phrenologist had a copy of Cinque's cast in his collection: H. D. Sweeter advertised that customers could view Cinque's cast at his office in Concord, New Hampshire, free of charge. Simultaneous with the December 1840 edition of the *American Phrenological Journal*, the Fowlers published an image of Cinque's cast in *The Phrenological Almanac for 1840*. The *Almanac* reprinted the most interesting or unusual profiles from the previous year's *Journal*. Images of Black Hawk, Aaron Burr, Maria Monk, and Antoine Le Blanc, "murderer of Judge Sayre, Morristown, N.J.," accompanied Eustache and Cinque, "leader of the Africans on board Schooner Amistad." Thus, viewers saw the "good Negro," Eustache, and the leader of a violent bid for freedom, Cinque, side by side.[122]

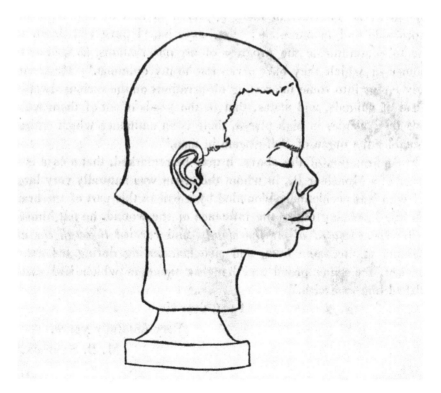

FIGURE 5.6. "Phrenological Developments of Joseph Cinquez, Alias Ginqua." L. N. Fowler, *American Phrenological Journal and Miscellany* 2, no. 3 (December 1839): 136. Courtesy of the American Antiquarian Society.

Fowler's cast of Cinque's head was one of several visual displays of the *Amistad* captives. John Warner Barber had already published *A History of the* Amistad *Captives* with a profile portrait of Cinque and silhouettes of the other Africans. Also during 1840, the artist Sidney Moultrap created life-size wax figures of the captives (complete with hair clipped from their heads), and Amasa Hewins painted a 135-foot-long panoramic scene of the moment of rebellion aboard the ship. In late 1840, Nathaniel Jocelyn painted Cinque's portrait for the Philadelphia Black abolitionist Robert Purvis.[123] Barber's drawing of Cinque in profile was based on pantograph silhouettes taken of all the captives. Cinque was the only individual Barber rendered in detail. Barber's portrait is a straightforward rendering of Cinque's likeness. It has facial details, but Cinque's head is presented against the blank page, just as the silhouettes of the other captives. Barber may have deliberately referenced physiognomy illustrations familiar to Americans. Absent from the profiles were color, eye shape, and facial expression. What remained as the focal point was the shape of the head, and the size and position of the nose, mouth, and chin—all emphasizing the African features of the sitters. In contrast to Barber's illustrations, Nathaniel Jocelyn's portrait of Cinque presented the African man in a context deliberately intended to provoke a sympathetic response in viewers. Jocelyn's Cinque was not dressed in the Western clothing he wore in the New Haven jail, as depicted in the *New York Morning Herald* engraving, Moultrap's wax figures, and Hewins's panorama. Instead, Cinque was garbed in a white toga-like garment, appearing "more like a Greco-Roman divinity than a brutish African marauder." Jocelyn's painting was deliberately timed to go on public display just weeks before the Supreme Court decision in January 1841.[124]

The Jocelyn portrait as well as the wax figures and Barber's silhouettes were in distinct contrast to most visual depictions of Blacks in the antebellum era. Edward Clay's popular "Life in Philadelphia" lithographs are more typical renderings: a series of vignettes show Black Philadelphians in middle-class finery shopping, singing, and going about their social activities. The images were clearly intended to ridicule Black men and women for imitating whites. Satire emphasized their exclusion from white society rather than suggesting that Blacks might share white middle-class values. The popular minstrel shows in the antebellum era also evoked laughter and ridicule in white audiences who viewed the spectacle of Blacks aspiring to white social and cultural behavior. Phrenology was one target of these performances: Sharpley's Minstrels and Ethiopian Burlesque Troupe performed "The laughable burlesque lecture, on animal magnetism, phrenology & physical knockings" for Philadelphia audiences, while in Sing Sing prison in New York, the

Croton Minstrels delivered "A Burlesque Lecture on Phrenology." A depiction of a Black phrenologist, titled "Free-Knowledgy, or Black Bumpology," in the *Crockett Comic Almanac* for 1839, echoed Clay's depictions of Black Philadelphians and blackface minstrel performances. Charles White's farce, "Wake up! William Henry: A Negro sketch known as Psychological experiments, Psychology, Bumps and limps, Bumpology, etc.," depicted Orson Fowler explaining phrenology to a Black man incapable of understanding Fowler's terminology. The explicit message of these comic performances was that phrenology was a ripe subject for humor. The implicit message was that Blacks, lacking the intelligence and education of whites, were fit subjects for phrenology, but not fit practitioners of the science.[125]

The *Amistad* captives continued to be displayed after their release from jail in January 1841. The *Amistad* Committee organized a tour of several northern towns and cities. Whether from curiosity or genuine interest and sympathy, the meetings were filled to capacity and beyond. As many as twenty-five hundred people attended at the Broadway Tabernacle in New York, where several of the *Amistad* men talked of their experiences and of their Christian conversion, and sang traditional Mende songs in their native language. This was the final public display of the captives but not the last time they appeared as phrenology subjects.[126]

Most of the *Amistad* captives returned to Sierra Leone in November 1841. Margru, known as Sarah Kinson, was one of the returnees. Sarah joined the new American mission in Komende. Five years later, she returned to the United States to study at the Oberlin Institute in Ohio. In 1849 she returned again to Sierra Leone to take up a post as the mistress of the Komende Mission girls' school. Sometime before her second return to Sierra Leone, she visited the Fowlers' office in New York City, where Orson Fowler did her reading. Fowler's analysis of Kinson appeared in the *American Phrenological Journal*, accompanied by a sketch of the now-adult Kinson (a pen-and-ink portrait of Kinson was created by William H. Townsend shortly after she arrived in the New Haven jail in 1839). Fowler related her history at Oberlin where Kinson "excelled in all branches of study and was one of the first scholars in the institution in mathematics and superior sciences." Fowler's phrenological comments were scant; he simply noted that "the forehead is large, sustained by a vigorous constitution. She is far superior to Africans generally." Once again, phrenology damned with faint praise.[127]

Twelve years later, the *Amistad* events still resonated enough with American readers for the *American Phrenological Journal* to once again show Cinque's cast, this time getting several facts wrong, including elevating Cinque to royal status. Side by side with this "prince of Africans" was the image of

Daniel Webster, "one of the great among the Anglo-Saxon race." Webster, according to the journal, was called the "Lion of the North," for his courage and determination. Fowler's *New Illustrated Self-Instructor* depicted Webster's "Lion Face." His prominent forehead and deep-set, piercing eyes indicated Webster's intelligence, fearlessness, and determination. The science of phrenology confirmed what Americans already knew: Cinque and Webster were superior examples of their respective races, but they were not, and never could be, equals.[128] Despite phrenology's prevalence in American culture, interest in the science waned quickly. By the end of the Civil War, phrenology lectures and publications were scarce. For those who studied the science of man, race continued to be the focus of experiment and speculation, although the tools and methods of researchers and writers no longer included phrenological evaluations. But for the few decades that the science held Americans' attention, it enabled them to articulate goals and ambitions, to criticize political opponents, advocate for reform, and argue about the place of Black men and women in white America.

CHAPTER 6

Fair America

Promoting American Invention

Agriculture, manufacturing, and all manner of domestic productions benefited from organizations that promoted and encouraged individuals to create, invent, and improve. Societies, institutes, and lyceums formed in the early republic fostered a constituency interested in science and technology. The decades-old American Philosophical Society for the Promotion of Useful Knowledge (APS) was a model for the societies created in the postwar era. Most of the APS members were merchants, scholars, or gentlemen—not scientists. And most activities of the APS related to practical and applied science, "such subjects as tend to the improvement of their country, and advancement of its interest and prosperity."[1] To achieve these goals in the early republic, natural resources had to be harnessed, a manufacturing base created, and the skills and knowledge necessary to carry out the vision of national development fostered. These endeavors went hand in hand with the creation of societies and institutions.

Even before the war was won, the necessity of developing natural resources and manufacturing goods in the United States spurred Bostonians to create the American Academy of Arts and Sciences (AAAS). The goals for the society were ambitious: "It is the part of a patriot-philosopher to pursue every hint—cultivate every enquiry, which may eventually tend to the security and welfare of his fellow citizens, the extension of their commerce, and the improvement of those arts, which adorn and embellish life."[2] The academy's charter enumerated the scope of the society's ambition:

> to promote and encourage the knowledge of the antiquities and the natural history of America; to determine the uses to which the various natural productions of the country may be applied; to promote and encourage medical discoveries, mathematical disquisitions, philosophical enquiries and experiments, astronomical, meteorological and geographical observations, and improvements in agriculture, arts, manufactures and commerce; and, in fine, to cultivate every art and science which may tend to advance the interest, honor, dignity, and happiness of a free, independent, and virtuous people.[3]

The academy understood the value of sharing knowledge and ideas. In stressing the importance of the mechanical arts and manufactures, for example, the AAAS hoped that "they who are engaged in these several branches of business may mutually aid each other." Everyone could contribute to "enrich and aggrandize these confederated States." Any and all "useful experiments and improvements" contributed to the "rising empire."[4] The first volume of the academy's *Memoirs*, published in 1785, illustrates the dual goals of promoting experimental and theoretical science and fostering practical applications for technology. Along with essays on astronomy, physics, and geology such as "Observations on Light, and the Waste of Matter in the Sun and fixt Stars," technological developments held a prominent place. Agriculture, too, received attention.[5] Articles from both the *Memoirs* of the AAAS and the *Transactions* of the APS were extracted and reprinted in the popular *Columbian Magazine* and the *American Museum*, proving that science and technology were of interest to an audience beyond these societies' memberships.[6]

The Academy of American Arts and Sciences and the American Philosophical Society encompassed any and all sciences and technologies. Smaller societies, some with equally ambitious agendas, were created in the immediate postwar years. Many indicated their goal to serve the nation's commercial interests by their name. *Useful* was a term many societies employed. The New York Society for the Promotion of the Useful Arts (1784), Trenton Society for Improvement in Useful Knowledge (1787), Kentucky Society for Promoting Useful Knowledge (1787), Alexandria Society for the Promotion of Useful Knowledge (1790), and the Society for the Promotion of Useful Arts (Albany, NY, 1792) were among the earliest to promote utility as their mission.[7] Most societies were short-lived. Some withered away from a lack of participation, others were absorbed into new associations. Ironically, many disappeared because, by the early nineteenth century, there were so many competing organizations devoted to improving the nation and its citizens.

Education was an integral part of this goal. The early nineteenth-century drive for knowledge spurred the creation of numerous organizations that sponsored lectures. In Boston, for example, Daniel Webster, John Lowell Jr., and others formed the Society for the Diffusion of Useful Knowledge in 1828. Lowell's interest in popular education led him to bequeath a substantial sum to fund a lecture series on the "philosophy, natural history, the arts and sciences, or any of them, as the trustees shall, from time to time, deem expedient for the promotion of the moral, and intellectual, and physical instruction or education of the citizens of Boston."[8] This generous funding guaranteed that the Lowell Lectures attracted luminaries, American and foreign, in a variety of disciplines. The fee paid to the Yale professor Benjamin Silliman for his twelve lectures on geology in 1836, for instance, was in excess of his yearly salary.[9] Silliman's lectures attracted between fifteen hundred and two thousand people per lecture—testimony to the interest in the nation's natural resources.

It was also a testament to the passion for improvement. Philip Hone of New York wrote in his diary in 1841 that "lectures are all the vogue." A writer in *Putman's Monthly* stated this more succinctly: "The Lyceum is the American theater."[10] Men and women formed local societies—often taking the name "lyceum"—that offered lectures to the local community. This movement achieved widespread participation and popularity because it fit well with existing modes of learning—aural and visual.[11] Science was entertaining as well as edifying. One New Englander recalled, "In Boston, all the boys and girls went to the Lowell Lectures. As many as five hundred at a time learned that acids were not alkalis and that Homer did not write the Iliad. It was quite the rage. The boys invited the girls, and after the lecture they walked home together, ending the evening with an oyster supper."[12] Lyceums and associations that sponsored lecturers in the second quarter of the nineteenth century created an important venue that exposed Americans to the intellectual building blocks required to develop and define American society.

Mechanics societies took a vocational approach to educating Americans in science and technology. The General Society of Mechanics and Tradesmen in New York City, for example, established a library and a school in 1820. The New York Mechanic and Scientific Institution's mission was instruction, but it also had a repository of machinery and tools available to its members in order "generally for enlarging the knowledge and improving the conditions of mechanics, artisans, and manufacturers."[13] Similarly, the Ohio Mechanics Institute aimed to make "the sciences, which have heretofore been taught only in our higher seminaries of learning . . . accessible to all who possess talent and taste to cultivate them." The institute offered lectures and classes to mechanics and apprentices.[14]

Agricultural societies, too, sponsored extensive outreach programs to educate farmers in new methods and equipment, with the goal of promoting the nation's independence from overseas products.[15] Agricultural societies were the first to use fairs as a means to reach a wider audience. The earliest was the Berkshire Agricultural Society's exhibition in 1802. When the Philadelphia Society for Promoting Agriculture held its first fair two decades later, one local newspaper reported attendance was estimated to be ten thousand—local farmers all interested in new crops, new fertilizers, and new inventions.[16] Fairs got people out and about—even if they did not read a science journal or attend a lyceum lecture. By appealing to "the patriotism, of every class in the community," fairs involved people as organizers, participants, and spectators.[17]

In the second quarter of the nineteenth century, the Franklin Institute of the State of Pennsylvania for the Promotion of the Mechanic Arts and the American Institute of the City of New York built on the momentum, practices, and popular interest generated by societies from the 1780s into the early 1820s. Both organizations strove to make the United States an economically independent nation with a significant role to play in world affairs. Technology was the key to achieving this goal. The largest and most successful of the early nineteenth-century societies, both organizations gave technology a public face—invention and innovation were not hidden away in workshops or shared within a small circle of like-minded enthusiasts. Both institutes strove to promote technology by exposing the American public to new developments, offering educational programs, and stimulating innovation through competitions. The most wide-reaching activities were the yearly fairs held by these organizations beginning in the late 1820s.

The Franklin Institute

The Franklin Institute was organized in 1824. In its first two years, the institute began lecture courses, a journal, an evening school, and a yearly technology fair.[18] The list of the institute's committees reveals the institute's aspirations for public engagement: Instruction, Library, Models, Minerals, Premiums and Exhibitions, Inventions, and Publications.[19] The contents of the *Franklin Journal and American Mechanics' Magazine* demonstrates how comprehensive the mission was and how broad a net the institute cast to support the interests of American mechanics, manufacturers, and inventors. The journal aimed to be "interesting to the artisan, and to the man of general reading." To accomplish this goal, the editor warned contributors that if articles were not written "in a style intelligible to the generality of readers, they will be in this respect altered; when unnecessarily prolix, abridged." The primary goal of the journal was "to diffuse information among artisans and manufacturers; and that it is therefore necessary to write in a style as familiar, and as little technical, as the nature of the subject will admit."[20] Topics included

> Mechanics and Natural Philosophy, Chemistry particularly in its application to the arts, American inventions and discoveries, whether patented or not, Internal Improvements, Natural History, Minerology, Botany, Mathematics, Architecture, Popular Education, Husbandry and Rural affairs [farming] particularly as regards the implements used, and the production of silk, flax, wool, cotton, dye-stuffs, and other articles employed in manufactures, Mechanical Jurisprudence, Foreign Journals, inventions, discoveries, and patents, Notices and Reviews of Publications relating to Arts and Manufactures, Miscellaneous articles, consisting of recipes, processes, &c.[21]

The journal was comprehensive and instructive. Every issue contained articles on practical instruction such as "the different methods of cutting screws; the preparation and use of varnishes and lacquers, of different kinds; the various modes of boring and drilling; the preparation and casting of plaster, and other materials."[22] The first five volumes included reports from the US Patent Office. The price of the *Franklin Journal and American Mechanics' Magazine*, four dollars a year, or fifty cents for a single issue, put it within reach of mechanics and artisans, a readership the journal hoped to attract.[23]

The Franklin Institute was also a site for education. Membership in the society gave individuals access to a reading room and library, cabinets of models, and lectures. The Committee on Instruction planned to start a high school. Though this did not materialize, the institute did offer instruction in both mathematics and drawing.[24] All these activities fostered opportunities for self-improvement among mechanics, artisans, manufacturers, and agriculturalists. The Franklin Institute's fairs expanded the audience for science and technology even further. Planned from the outset as an integral part of the organization, the fairs were intended to promote American products and technologies while simultaneously showcasing the talents, ideas, and results of American craftsmen and inventors. In concert with the classes, lectures, and publications, the fairs fostered widespread interest in American products and improvements. As the number of participants and spectators grew year by year, the institute sought larger venues to accommodate the numerous entries submitted and the crowds who came to see them.

The fairs were accessible to a great number of people because they were located in the central part of Philadelphia. The first exhibition, in 1824, was held in Carpenter's Hall. Located on Chestnut Street between Third and Fourth Streets, the building was in the city's central business district—close to transportation, hotels, businesses, and other cultural organizations.[25] The following year, 1825, the institute fair moved to the Masonic Hall on Chestnut Street between Seventh and Eighth, a building that accommodated a larger number of exhibitors and visitors. Between fifteen and twenty thousand people attended the fair in its first year. By the third year, there were thirty-four thousand visitors.[26] The length of the fair also increased as its popularity grew. Originally a three-day event, by 1838 it had developed into a weeklong affair. By the 1840s, the institute's fairs were so well known that they were noted in guidebooks to the city.[27]

Fairs helped the Franklin Institute achieve several of its goals. One was to make "producers and purchasers acquainted with each other."[28] In other words, the fair was an advertisement for American products. Moreover, consumer acquaintance with domestically produced goods drove home the fact that American producers aimed to make their goods competitive with imported ones; purchasers were assured that the products on display could be had "on as good, or better terms than elsewhere."[29] Producers were given incentives to participate in the fairs. Gold, silver, and bronze prizes were awarded for a selection of high-quality domestically produced goods as well as technical improvements to machines and devices. The prizes were intended to "promote the industry, skill, ingenuity, and enterprise of the country."[30] All types of goods received attention from the prize committee: hats and bonnets, textiles, ceramics, and paint, steel, and cast iron. Products were required to be made in the United States from domestic materials.[31] But domestic production was not sufficient to earn merit from the prize committee. The institute emphasized that the quality of entries must equal or better imported goods. A silver medal for the best cotton cloth, for example, required that the textile be "of superfine quality, in imitation of English Cambric Muslin."[32] In 1826 George Armitage's plated mugs met these exacting standards. The committee deemed the mugs "equal in taste and design, and in workmanship, to the best imported ware."[33]

There was attention to all aspects of production and trade. Silver medals were awarded for the invention of "an apparatus practically superior to any now in use, for heaving up a Ship's Anchor"; for "a method better than any in use, to protect timber in ships, or other works, against the effects of the dry rot"; and for the construction of "a Marine Railway, for hauling ships out of the water."[34] The institute placed special emphasis on coal-related products and appliances. There were several prizes for anthracite-burning devices: a silver medal for "the best constructed Furnace, for consuming anthracite in generating steam, to be applied to steam engines," and a gold medal "To the maker of the greatest quantity of Glassware, not less than 100 lbs. The fuel used in the manufacture to be not less than 3–4 anthracite coal," and to "the person who shall have manufactured in Pennsylvania, the greatest quantity of Iron from the ore, using no other fuel but anthracite, during the year ending September 1, 1826. The quantity not to be less than twenty tons."[35] Domestic as well as industrial applications were encouraged. A silver medal was offered for an anthracite-burning kitchen stove that would "unite convenience with economy." The criteria considered several aspects of what consumers expected from appliances: efficiency, low cost, and "tastefulness of design."[36]

Though the vast majority of prizes for machines and goods were awarded to men, several women won awards for straw bonnets. Woven in imitation of bonnets imported from Italy, they were a popular accessory in the early nineteenth century. Hannah Smith, a teacher at the Walnut Street school in Philadelphia, and Eliza Bennet, of Massachusetts, both won awards for bonnets that were "very superior, and deserving particular notice."[37] The institute acknowledged the work of American artists as well as artisans. In 1832 the engraver J. Hill won a prize for his aquatint of Thomas Doughty's oil painting of the Fairmount Water Works—an appropriate subject for an industrial exhibition.[38]

Everything in the institute's fairs was of American production using American materials. The awards clearly conveyed the message that developing American ingenuity and harvesting American resources was paramount. Speeches reinforced this imperative. The civil engineer Solomon W. Roberts's closing address at the institute's sixteenth fair in 1846 was especially strident; Roberts credited the institute with rousing "that patriotic spirit which has already dawned on our countrymen, to rally them around the American standard, and free them from the shackles of the foreign mechanic." Roberts reminded his audience that Britain was the United States' greatest rival. As a free and independent nation, the United States could not only compete with Britain but excel that nation in the production of essential products such as iron and coal: "Why should not Philadelphia excel Liverpool and Manchester? and why should not Pittsburg go beyond Birmingham and Sheffield. They may yet do so if their natural advantages are fully developed." Encouraging domestic industry was "a national object, which we are called on by every dictate of true patriotism to promote." Echoing Benjamin Franklin, the spirit behind the institute, Roberts told his listeners that "as it is with an individual, as it is with a family, so it is with a state, and so it is with a nation. Dependence upon others, lack of industry and enterprise, want of variety of occupation, and habits of extravagance are the high roads to poverty: while a due sense of the dignity of labor, persevering self culture, faithful application, and prudent economy lead to competence and independence and often to wealth."[39] At the 1828 exhibition, American-made machinery printed copies of patriotic verse. James M'Henry's "Ode," written for the exhibition, drove home the importance of American artisans:

> The envy of nations by despots controll'd
> Who can witness no charms such as here we behold;
> While Invention's bright wand,
> In the artisan's hand,
> Points to glory the freemen of Freedom's own land.[40]

Visitors could not ignore the message: American production was an expression of patriotism and national pride.

The Franklin Institute's annual fairs were attended by thousands of men and women, the majority of whom were not specialists, mechanics, or manufacturers. Visitors saw working models of machinery and appliances, textiles, tools, and apparel. There was something for everyone, from mechanics and inventors who entered their items for competition, to individuals looking to purchase better tools, apparatuses or machines, to consumers in the market for domestic goods such as stoves, lamps, bromides, or umbrellas. Promotion was clothed in the rhetoric of nationalism—the ambition to make the United States competitive with Britain by producing American products as good or better than foreign ones. This goal was carried further into the realm of political economy by the Franklin Institute's sister organization, the American Institute of the City of New York.

The American Institute

Four years after the Franklin Institute was founded, a group of businessmen in New York City incorporated the American Institute of the City of New York. Like the Franklin Institute, the American Institute was devoted to promoting American invention and production. It also used patriotism to emphasize the United States' need to challenge Britain for supremacy in domestic trade. Unlike the Philadelphia organization, the American Institute sought government assistance in its endeavors by lobbying for protective tariffs. The institute made the campaign for tariffs the centerpiece of its annual fairs in order to show the public in a very concrete way that protective tariffs benefited not only American producers, but American consumers too. The American Institute published a journal and established a library for its members. But, unlike the Franklin Institute, which sponsored educational programs and lecture series, the American Institute used its annual fair as the primary way the organization interacted with the general public.[41]

From the outset, the American Institute's fairs surpassed those of the Franklin Institute. There were more entries and more visitors. By the time of the third fair, in 1830, between thirty and fifty thousand spectators were attending the event.[42] The fairs repeatedly outgrew their venues; first the Masonic Hall, then Castle Garden, then Niblo's Garden. In 1855 the institute rented the Crystal Palace—the largest public venue in the United States. The popularity of the American Institute fairs owed much to location; by the 1820s New York City was a rapidly growing metropolis with the largest port in the United States. Commerce was the heart of the city.

The scope and variety of the entries at the fairs guaranteed something for everyone. The institute was organized into four departments: Agriculture, Commerce, Manufacturing, and the Arts. All four were represented by entries and prizes. Artisans, mechanics, manufacturers, and agriculturalists exhibited their inventions and improvements. Art and crafts were also highlighted: painting, sculpture, and handiwork of all sorts found an appreciative audience. The scope of the exhibits represented a large portion of the American public. This strategy of inclusiveness encouraged domestic productions and at the same time invested makers with a sense of the institute's mission to promote and protect American products. At the fair in 1830, for example, a prize was awarded to a model called the "national Plough."[43] Premiums were given to items large and small. In addition to textiles, there were nails and stoves, glass and earthenware, paper and stationery (including sealing wax, ink, and pens), hats, boots and shoes, miscellaneous articles such as shell combs, snuff, razors and penknives, pianos, cabinets, pictures and sculptures, chemicals (such as carmine made from cochineal), and machines and models of all kinds. The institute report for the 1830 fair proudly asserted, "In *cut glass*, buttons, scales, silver-plate . . . ladies' and gentlemen's hats, shoes and boots, paper &C. we fear no foreign rivals."[44] As the many mechanical models illustrated, the American Institute was eager to showcase improvements in technology and to highlight new inventions and processes. The twenty-sixth annual fair, in 1853, for example, included a premium for daguerreotypes.[45]

The institute encouraged American production of all kinds, including household goods. The prize committee offered awards to encourage individuals, especially women, so that "the good effects of our Institute should be carried to the firesides of our fellow-citizens, an impulse communicated to real home industry, and the exertions and handy work of our fair countrywomen be particularly stimulated and promoted."[46] The premiums for "Articles of Domestic Produce and Manufacture" included hearth rugs, silk and linen thread, quilts, beadwork bags, and scrap tables. In 1830 Mrs. Joshua Stow of Middletown, Connecticut, received an award for "several pounds of fine Linen Thread, spun by her in her 69th year." And an unnamed young lady in Connecticut, received a prize for merino wool stockings, "the yarn spun by herself." Reminiscent of the encouragement of home industries during the Revolution when British products were boycotted, a call to patriotism inspired a new generation of women to work for the national good.[47]

To drive home the institute's emphasis on the need for protective tariffs, the *Catalogue of Articles*, available to visitors as they toured the fair, listed items, the makers, and if an import tariff had existed, the duty that would have been owed for an import of the same kind. Miss Conklin's patchwork quilt, James Warner's microscope, Nathan P. Ames's carpenter's tools, "equal to the best English," and the mill, crosscut, and wood saws, exhibited by Verree and Company of Philadelphia (also "superior to English") were all listed with a 25 percent duty charge for imports of similar make and kind. M. Haddock's innovative "machines propelled by mice" merited a 30 percent import duty. The artist William Sidney Mount's painting "Rustic Dance" and E. Bryan's "Incorruptible" imitation teeth were given a 15 percent duty equivalent.[48] The message conveyed by drawing attention to potential duties was clear: replace imports with domestically produced goods.[49] Textiles, silk, wool, and especially cotton received special recognition: "*Cotton prints* in fineness, delicacy, and richness of colour, have greatly surpassed all former exhibitions; and other cotton goods have arrived nearly to perfection . . . manifesting the beneficial effects of increasing duties, and strengthening the policy of protecting home industry."[50] Cotton production was crucial to the national welfare. This dependency of northern manufacturers on southern producers was not lost on the institute. No one could ignore the growing tensions between southern agriculturalists and northern manufacturers: a toast given at the annual dinner that year was to "Our brethren of the South—The cotton which they transmit from their rich and ever productive soil, we return to them in useful fabrics from our spinning-wheels and looms."[51] In the era of escalating regional conflict over tariffs, this attempt to mollify southerners, though well-intentioned, was merely rhetorical. The institute firmly supported any and all tariffs that protected American products from foreign competition.

The American Institute celebrated the success of each fair with speeches, toasts, and songs. At the tenth fair dinner, held at Knickerbocker Hall in October 1837, the Sacred Music Society sang an ode composed for the occasion:

> Awake to song and loud rejoicing,
> Strike every chord with bounding hand;
> Come sing of Genius' enterprising,
> For here she finds her foster-land.[52]

A second ode, sung by Captain Joseph Cowdin, celebrated the inventor Robert Fulton: "For as long as the sun in his splendour is seen, / His name shall endure, and his laurels be green."[53]

Toasts to scientific achievements and technological innovations abounded at the annual dinners. Electromagnetism was celebrated as "a science which promises to subject the artillery of heaven to the will of man. When rail-road cars and steam-boats are moved by lightning, and saw-mills by thunderbolts, then indeed we shall have arrived at the Ultima Thule of the mechanic arts."[54] Railroads, canals, and steamboats were acclaimed as "potent instruments in the hands of a free people."[55] The opening and closing ceremony speeches by prominent invited guests lauded the institute for its efforts to promote American products and American manufacturing by rewarding innovation and industry. Congressman Edward Everett of Massachusetts, an ardent supporter of the American System—a national bank, protective tariffs, and federal support for infrastructure—praised the institute for sponsoring fairs that enabled the organization "to awaken, to guide, to stimulate, and to reward the industry of your fellow citizens, by this beautiful and commodious display of its products, and by the honorary premiums, which you have assigned to some of its most successful efforts."[56] Congressman Tristram Burges of Rhode Island assured his listeners at the 1830 fair that the United States was well on its way to fulfilling the ambitions behind the American System: to "draw from foreign countries no other products than those, which cannot be had from our own labor, or be grown under our climate, on our own soil; . . . and by sending abroad our manufactures, render foreign nations our ultimate consumers, and thereby relieve ourselves from the cost of making exchanges with them." Burges described the institute's efforts in the sphere of political economy as acts of "untiring patriotism."[57] One contributor to the institute's journal asserted that, by means of the fairs, "the power, wealth, and resources of the country, more than by any other means within my knowledge, are destined to be developed."[58] There was much self-congratulation in these toasts, speeches, and comments by the institute's supporters. They articulated shared ideas about political economy, the value of labor, and patriotism, all critically important for the nation to meet the dual challenges of foreign competition and increasing domestic conflict.

The Crystal Palace

Almost thirty years of industrial and agricultural fairs prepared Americans for the grandest fair to date—the Exhibition of the Industry of All Nations, held in New York City from 1853 to 1854. This fair surpassed the Franklin Institute and American Institute events in several ways. The exhibition was far larger than any of the institute fairs; it accommodated far more exhibitors and visitors. It was international. And it was a perpetual fair—lasting almost continuously from July 1853 until November 1854. Most importantly, the exhibition was held in a purpose-built venue: the Crystal Palace. Aided by continuous news reports and promotional advertising, the Crystal Palace caught, and held, the nation's attention from the groundbreaking ceremonies in 1852 until its fiery demise six years later.

The Exhibition of the Industry of All Nations in New York City was a direct result of the Great Exhibition of the Works of Industry of All Nations at the Crystal Palace in London. The British exhibition, held in 1851, displayed machinery, manufactured goods, and art from around the world. It was housed in a building that was itself a technological achievement of steel and glass.[59] The exhibition opened on May 1, 1851, and closed in October. In those five months, six million people visited the Crystal Palace. Twenty-five nations contributed displays to the exhibition. Half the space was reserved for British products, with the other half shared by the rest of the contributors. The exhibition was organized by country, making it easy to recognize a nation's contributions. The United States was well represented, both by manufactured goods, inventions, and art.[60] Five hundred and forty-one American items were on display: machines (including the cotton gin), manufactured goods, agricultural products (corn, cotton, iron ores, and minerals) as well as food and drink (biscuits, preserved peaches, and sarsaparilla) showcased the country's natural resources and its citizens' ingenuity.[61] Agricultural machinery, which the United States had developed to a much higher standard than Britain, was particularly well-represented. Though the British press was initially critical of the American displays, demonstrations of agricultural tools, especially plows, drew attention and admiration. The British humor magazine *Punch* conceded that the Americans impressed visitors with their ingenuity and technological innovations. Adding new verses to "Yankee Doodle," the magazine celebrated the United States' triumphs:

> Yankee Doodle went to town
> His goods for exhibition;
> Everybody ran him down
> And laughed at his position;
> They thought him all the world behind
> A goney, muff, or noodle.
> Laugh on, good people—never mind—
> Says quiet Yankee Doodle.
> Chorus—Yankee Doodle, etc.
>
> Your gunsmiths of their skill may crack
> But that again don't mention;
> I guess that Colt's revolvers, whack
> Their very first invention.
> By Yankee Doodle, too, you're beat
> Downright in agriculture,
> With his machine for reaping wheat,
> Chaw'd up as by a vulture.[62]

American inventors won several Council Medals, the highest award given: Cyrus McCormick for his reaping machine, Charles Goodyear for India rubber, and Gail Borden Jr. for his "patent meat biscuit."[63]

By most measures, the United States had a successful fair. In Massachusetts, the *Springfield Republican* credited the achievement to the American system of government: "We have demonstrated that a nation choosing its own rulers, and governing itself, is best adapted to progress in art and power."[64] Charles Rodgers, who vigorously promoted and defended America's superiority, boasted, "The most important inventions, those which confer the greatest amount of power on mankind, in the ways of industry, were unquestionably furnished by American citizens. In this opinion, the entire world concurs."[65] Despite the prizes awarded and the public accolades, Rodgers believed the London Exhibition did not do justice to American products: "Probably no American who saw our contributions in London, who did not feel some regret that they were not a more just and equally successful exponent of our resources, industry, and arts."[66]

The *New York Tribune* editor Horace Greeley agreed: "Our share in the Exhibition was creditable to us as a nation not yet a century Old . . . but it fell far short of what it might have been and did not fairly exhibit the progress and present condition of the Useful Arts in this country. We can and must do better next time."[67] The *Cleveland Plain Dealer* took a stronger line. It declared rather belligerently, "The English seem desirous of being further beaten. Let us not deny them the boon!"[68]

In 1853 a group of investors took up Greeley's call for the United States to do better next time. They formed the Association for the Exhibition of the Industry of All Nations. Their confidence in the success of an American exhibition was bolstered by the widespread press given to the British exhibition, the frequent newspaper articles, books, and even local attractions such as a diorama of the London Crystal Palace displayed at the Stuyvesant Institute in New York City.[69] Moreover, an American fair provided an opportunity for the United States to demonstrate the material benefits of republican government. The *Cleveland Plain Dealer*, always a booster for American supremacy, asserted that "no country is better fitted than the United States to give the next impetus to this great idea of modern civilization. Its effect on our interests abroad and at home will be cheering, indeed. We can point out to the intelligent foreigner the advantages of republican institutions and easily remove the film of prejudice that has so long beclouded his western gaze."[70] Despite such flag-waving supporters, unlike the London exhibition, the New York exhibition was a private endeavor. There was no government subsidy to build the venue or to aid with running the exhibition. The association undertook every aspect of the enterprise, from initial planning to soliciting exhibitors and staging the opening ceremonies. The association was proud of its achievement but also dismayed at the federal and state governments' refusal to subsidize the venture.

The New York exhibition was deliberately modeled on the one in London. The British had hit on a winning formula: central location, numerous and varied exhibits, and a purpose-built, state-of-the-art venue. The Americans saw no reason to stray from this template. The association determined to capitalize on the name of the building as well as on the exhibition's format. Yet the New York exhibition's champions insisted that their Crystal Palace was not only different but superior to London's. As the building began to rise on Reservoir Square (42nd Street between Fifth and Sixth Avenues), promotion for the exhibition proliferated in daily newspaper reports on construction progress. Monthly magazines of all sorts—from *Harper's Weekly* to *Godey's Lady's Book* as well as those devoted to science and technology, such as *Scientific American* and Benjamin Silliman's *World of Science, Art and Industry*—kept readers informed. *The Plough, the Loom, and the Anvil* told its readers that the exhibition would showcase and celebrate the efforts of American workers: "There will be gathered here the choicest products of the luxury of the Old World, and the most cunning devices of the ingenuity of the New. The interests of Manufacture, Commerce, and the Arts, will all find encouragement and protection within these walls. . . . Here will be collected multitudes of all nations; but the great and crowning feature of the enterprises, that it will offer amusement and recreation to the working classes, such as they can find nowhere else; that it will be a PALACE FOR THE PEOPLE."[71] Even children's publications like *Merry's Museum and Parley's Magazine* praised the New York Crystal Palace: "It will be the largest and most beautiful building ever constructed in this country, and for architectural effect and convenience of arrangement for purposes of exhibition will be far superior to the London Crystal Palace."[72]

The January 1853 issue of the *National Magazine; Devoted to Literature, Art, and Religion* provided an illustration of the exterior and the floor plan. The magazine told readers that the Crystal Palace would be "the largest edifice ever put up in this country." It conceded that the New York Crystal Palace was smaller than London's, but the New York building compared favorably with Old World monuments such as St. Paul's Cathedral in London (which was "decidedly smaller") and St. Peter's Basilica in Rome. The latter church was larger but only just. The magazine, perhaps carried away by American boosterism, stretched the truth when it told readers that the Crystal Palace was an American design. "The beauty of the structure is manifest at a glance. It is an honor to the nation; and the more so, as it is not constructed from foreign designs. . . . 'The directors,' says an exchange paper, 'have been fortunate in selecting a plan from this side of the water, and in not going to England for one.'" Conveniently, the magazine overlooked the fact that neither of the architects, Georg Carstensen and Karl Gildemeister, was American. Enthusiasm and nationalism carried the day: "An industrial display will be made within its walls, such as shall be creditable not only to the country, but the age."[73]

The exhibition opened in July 1853 with fanfare. The imposing new building was filled with goods, machines, art, and crafts from around the world. The New York exhibition was not as large as London's nor were as many nations represented. Nevertheless, the exhibition fulfilled the ambition of the organizers to showcase American products and American ingenuity.

The opening ceremonies reinforced the national importance of the exhibition. The president of the United States, Franklin Pierce, was the guest of honor. Cabinet members, governors, and legislators were also part of the celebration. Numerous speeches reiterated the exhibition's significance as a symbol of national progress. President Pierce said that it was designed to "promote all that belongs to the interest of our country." The exhibition fulfilled the important mission of "strengthening and perpetuating that blessed Union."[74]

Not everyone agreed with Pierce's assessment. Secretary of War Jefferson Davis, a southerner and adamant opponent of protective tariffs, disrupted the harmony President Pierce envisioned among the states with his remark that free trade was a necessity. Other southerners did not share the enthusiasm for the exhibition. The North Carolina *Wilmington Journal* complained, "Under the specious pretense of an American edition of the World's Fair, a company of New York speculators are palming on the country the vilest and most contemptible humbug even of this age of humbugs. The New York Crystal Palace in its conception and management would do no discredit to the genius of Barnum." Nor did the *Journal* agree that the New York Exhibition was superior to London's. On behalf of southerners who made the long, expensive journey north to see the Crystal Palace, the *Journal* stated: "We have been caught. Thousands of inquisitive people from the South have travelled to New York through dust and heat; have suffered the torments of an over-crowded city, have gone to the Crystal Palace, had their pockets picked, their persons searched, their toes mutilated, and for what?"[75]

What visitors to the Crystal Palace did encounter were the sights, sounds, and even smells redolent of progress. The building itself was a testament to the cutting-edge engineering responsible for the soaring height of the interior, the open spaces, and the abundant natural light from the glass windows encasing the steel-framed structure. Visitors were reminded of great Americans past and present. In the very center of the building, an equestrian statue of George Washington dominated the central axis. Further down one aisle was a bust of the Massachusetts congressman and tariff supporter Daniel Webster.[76] In the Machine Arcade, working steam engines provided the motor power for many other machines throughout the exhibition. Models and full-size machines whizzed, clanked, and puffed throughout the day.[77] Other sections displayed fragrant coffee and spices; perfumes and soaps scented the aisles where manufacturers displayed their products.

Newspapers and magazines prepared visitors for the exhibition and suggested which displays they would most want to see.[78] The official guidebook, sold at the entrance, impartially directed visitors to each section of the exhibition—art, crafts, industrial machinery, manufactured goods, and agricultural products. One unofficial guidebook, *A Day in the Crystal Palace and How to Gain the Most of It*, by William Richards, promoted itself with the argument that the "Official Catalogue, indispensable in itself as a complete and systematic inventory of the thousands of objects embraced in the Great Exhibition, is, yet, in the very nature of the case, deficient in that sort of information concerning the chief attractions of the Palace which the visitor requires."[79] Richards's tour was more selective: he gave his readers a route to walk through the building and told them which items were worth viewing along the way. Richards began with the building itself: he compared the dome, "the largest dome in the Western World," to a balloon "expanded and impatient for a flight into the far-off sky."[80] Although the agricultural machinery did not "possess sufficient interest to the general visitor," national pride made it worth pointing out "the now world-renowned machine of McCormick—the great Illinois reaper, which opened the eyes of our excellent neighbour, John Bull, to the genius and energy of Yankee farmers."[81] The machine arcade, on the other hand, could not be missed. Richards asked his readers to admire the American-made steam engines, "all together in motion; the huge fly wheels revolving, and the polished arms, levers and beams reciprocating with the very sublimest 'poetry of motion' our imagination can comprehend." Further along, the spinning machines and looms "whirl, and whiz, and thump, and bang, with delightful unanimity."[82] Richards was a good guide: the machine arcade was one of the most popular sections. In its second year, the newly reorganized exhibition included even more working machinery and demonstrations. The managers promised that visitors would see "operating specimens of nearly every great invention, and in some instances the entire process of manufacturing various fabrics will be exhibited."[83]

The exhibition was also an extravagant advertising space, a point not lost on *A Day in the Crystal Palace*: "Small manufactures appear in abundance as we proceed along the passage. Sand-papers, buttons, wooden-boxes, and wash-boards alternate with each other, and with clothing apparel, the latter so abundantly displayed, that we may reasonably consider the gallery a grand furnishing magazine!"[84] The exhibition catered to the increasing demands of consumers for clothing, furniture, home decoration, kitchen appliances, and toiletries. Eye-catching displays drew visitors' attention to particular vendors. Even soap was displayed in a unique and attention-grabbing way: "Soap rises in pyramids and columns, like monuments in a cemetery. The most unique exhibition of it, however, is made, by Taylor & Co. of Philadelphia, in the form of a soap window, bearing a deceptive resemblance to stained glass."[85] During and after the Crystal Palace Exhibition, manufacturers capitalized on the popularity of the fair. Two piano makers, T. H. Chambers and Hazleton and Brothers advertised "Crystal Palace Premium Piano Fortes." J. M. Dow and Company of Barre, Massachusetts, offered customers "Crystal Palace" dress hats. John A. Powers in Milford, New Hampshire, sold "Crystal Palace" cooking stoves. Lucy Guild, of Rupert, Vermont, sold a Crystal Palace Chart for Cutting Dresses. She warned customers not to accept imitations: "none genuine unless the Crystal Palace is engraved upon it." Advertisements by Gravesbeen and Company Pianos in the *Washington Sentinel* and Sevres China in the *New York Herald* included the information that they had exhibited at the Crystal Palace. Wheeler and Wilson's advertisement for their sewing machines proudly announced the "Highest Premiums again awarded by the *American Institute*, Crystal Palace, *Maryland Institute*, Baltimore, and at the *Maine, Connecticut*, and *Illinois State Fairs*."[86]

American Accomplishments

The material abundance that visitors experienced reinforced a key message implicit in the design and organization of the exhibition. Displays were presented by country of origin, and the United States contributed the lion's share of items in every category except art.[87] As the poem "Crystal Palace and the World's Fair in New York City" declared,

> The great Crystal Palace which made such a show,
> Was in London, in the old world, where millions did go,
> But now in the new world their example we take,
> In hopes that our Yankees good profit will make. . . .

> Mechanics and merchants and professional men,
> With free navigation have a chance now and then;
> Still Yankee invention keeps pace with the age,
> And they are sure to succeed if but once they engage.[88]

Unlike the American Institute's direct methods, the work of shaping the Crystal Palace Exhibition into an articulation of national pride was left to the journalists. In contrast to the official catalog visitors carried with them through the exhibition, later published catalogs, owing to their size, weight, and cost, were meant to be souvenirs rather than guidebooks.[89] With ample space to include commentary as well as illustrations, these books enlarged on the theme of national superiority. The same was true of the *New York*

Tribune's coverage of the exhibition; lengthy and frequent articles about various aspects of the exhibition appeared in the newspaper from early in 1853 into November, when the exhibition closed for the year. The editor Horace Greeley, an early champion of an American exhibition, later compiled and edited the Tribune articles as *Art and Industry as Represented at the New York Crystal Palace*. Greeley praised the superiority of the New York Crystal Palace over the London building: "But the lofty, magnificent dome of the American Palace has no parallel in the British, and probably none in the world unless it be that of St Peter's in Rome. . . . Its galleries are relatively finer and more spacious. The eye is better satisfied with its symmetry, its decorations, and its colors."[90] For Greeley, the New York exhibition was a testament to the United States' growing power: "In this we behold the first decided stand of America among the industrial and artistic nations of the earth. In this we see a recognition of her progress, power, and possibilities."[91] Greeley argued that the Crystal Palace should become a permanent institution "whereon genius or ingenuity may at once place its productions and obtain the highest sanctions."[92] No existing organization could achieve the national prominence that the Crystal Palace had already demonstrated. Not even the two great institutes that inspired the Exhibition of the Industry of All Nations could hope to compete with the Crystal Palace; neither the Franklin Institute nor the American Institute could serve as a national showcase for American ingenuity and talent. The editor of *Scientific American* echoed Greeley's enthusiasm. American citizens should feel proud of the nation's accomplishments: "John Bull pointed sneeringly at the empty space of the American department in the London Fair: but although we had little interest in advertising our wares among them, while America is their great market, the tables are completely turned." Britain could not compete with American inventions: "Although they [the British] exhibit many things of great artistic merit, who can point out to us a Hobb's lock, a McCormick's reaper or the model of an *America?*"[93] *Scientific American* continued to promote the technological prowess of the United States; from early 1853 until the exhibition closed in the fall of 1854, the magazine ran a weekly column, with a masthead image of the Crystal Palace, devoted to reports on the exhibits. The editor selected various departments, primarily the machines and mechanical processes, for detailed analysis and commentary.

Another compilation of the exhibition, Charles R. Goodrich's *Science and Mechanism: Illustrated by Examples in the New York Exhibition, 1853–4* contained lavish illustrations of the many machines and inventions on display. Goodrich devoted particular attention to the technologies and products that were important to the American economy. His frequent commentary reinforced the significance of the United States' contribution to technological development. An American-designed ordnance machine, for example, not only exhibited "the perfection of mechanical contrivance, but even the British Parliament considered it "as an authoritative precedent for similar establishments in England."[94] Goodrich was notably effusive in his praise of agricultural and horticultural machines as examples of "the mechanical genius of our countrymen." Goodrich deemed these machines, especially the reapers, so advanced that they "appear to have reached the limits of economy, efficiency, and convenience, and to be susceptible of no further improvement."[95] Even American paper was superior: exceeding "in fineness and smoothness the heavy English paper a few years ago regarded as indispensable."[96]

Goodrich especially commended American productivity in textile manufacturing: "Its machinery has been brought to such perfection, and its resources so far developed, that it is quite independent of tariffs, and able to compete successfully with the industry of any other country. Its rapid growth and magnitude are unparalleled, except by the rise of the same manufacture in Great Britain." Woolen and worsted fabrics, too, were "an important branch of national industry." Flax and hemp were singled out as a "new and highly important branch of American industry." Even dyed and printed fabrics were competitive with imported ones: "The lawns, calicoes, mousseline de laines, &c., shown by the print-works of Massachusetts, Rhode Island, and New York, will compare favorably with similar productions from any country." Even technologies that were not of US creation benefited from American talent. The daguerreotype, Goodrich told his readers, "in America, has arrived at great perfection, which is in a great measure due to the extreme clearness of the atmosphere; aided, however, by skillful manipulation. The pictures exhibited are remarkable for a brightness and distinctness observable in no other country."[97]

The exhibition continued into the early fall months. But, by then, many exhibitors had withdrawn their items, and attendance had dropped precipitously. George Templeton Strong presciently commented, "Crystal Palace losing money fast according to Dupont, who ought to know. It will prove a bad speculation, unless they make P.T. Barnum chief director."[98] *Scientific American* declared that the exhibition was "now dying a natural death." The editor ruefully noted the deserted space: "A stranger visiting the Palace for the first time and reading the regulation, 'The Police will require visitors to move along when the passages become crowded,' would be very much disposed to enjoy a laugh at the expense of the enterprise."[99] The exhibition limped along, but it did not close. Theodore Sedgwick resigned as head of the association, and Phineas T. Barnum was appointed in his place. Barnum was the obvious choice to reinvigorate the faltering exhibition: he had a proven ability to attract crowds. The popular women's magazine *Godey's Lady's Book* applauded his appointment: "The great Barnum is elected president of the Crystal Palace Association in New York. There is now some hope of its success. There was none under the former administration, for a more decided old fogy concern we never heard of—kid-gloved gentry, who had about as good an idea of managing an establishment like the Crystal Palace as they had of earning the money which their fathers left them."[100]

Barnum's American Museum had been a successful entertainment venue for over a decade. As William Withington observed in his poem "Crystal Palace and the World's Fair in New York City":

> Remember friend Barnum he keeps in Broadway,
> A large splendid museum by night and by day,
> Before you leave New York be sure you go there,
> For you have to pass that as you go to the fair.

Barnum's strategy at the museum was to continuously supply new entertainments alongside the permanent displays. Natural history rubbed shoulders with fabrications such as the Feegee mermaid. Moreover, Barnum knew his audience. Barnum put new technologies on display next to singers, dancers, singing dogs, balloon ascensions, exotic animals, and unusual humans such as Tom Thumb and the Siamese twins Chan and Eng. In 1843 he advertised that a room at the museum would house a perpetual fair at which Barnum encouraged "mechanics, Tradesmen, Manufacturers, Inventors, Artists, and all men of business" to display their items. Barnum took care to mention to potential exhibitors that this strategy was "the most efficient and cheap mode of advertising," especially since Barnum employed a man to "point out the merits of the articles deposited, make sales, and obtain orders on Commission." Here, visitors saw working models of machinery such as a rotary knitting loom. In typical Barnum fashion, the loom was alternately powered by a miniature steam engine and a dog.[101]

Barnum and the association reinvigorated the Crystal Palace exhibition with new displays, more entertainments, and a lower admission price. Prizes were announced in newspapers throughout the nation. A gold medal worth one thousand dollars and four silver medals of lesser value would be awarded for "the most useful and valuable Invention or Discovery which shall have been patented or entered into the United States Patent Office during the year closing the first day of December next, provided only that the said Invention or Discovery, by specimen model or product, shall have meantime been exhibited in the Crystal Palace." This policy accomplished several goals: it promoted American ingenuity, protected the intellectual property of the inventor, and showcased the exhibition as an organization that fostered innovation.[102]

In preparation for the revised and improved exhibition, the Crystal Palace closed for a month in April 1854. When it reopened in May, Barnum staged a grand ceremony. No president, governors, or legislators were in attendance this time. Instead, the ceremony honored the workingman. Barnum gave a short speech. Horace Greeley, editor of the *New York Tribune*, and Henry Ward Beecher were two of the celebrity orators. All the speakers emphasized the Crystal Palace as a "Palace of the people," where the products of labor and labor itself were celebrated.[103] As an ode composed for the exhibition emphasized,

All nations flock unto the Mart,
That's held on Freedom's Soil,
Contending for the golden prize,
That's gained by honest toil.[104]

Ignoring the international character of the exhibition, many of the speeches celebrated it as a demonstration of American achievement and American promise. Barnum told his listeners "We hope to bring forth our new race of heroes . . . conquerors upon the battle-field of labor—victors in the sublime struggle of handicraft and intellect with ignorance and inertia. We hope to make such heroes of you, industrials, who listen to me—to immortalize you in the immortalization of our age and nation."[105]

Barnum planned the rejuvenated Crystal Palace as an entertainment venue as well as a showcase for technology and design. Just as at Barnum's Museum, the Crystal Palace offered daily concerts. Newspaper advertisements for Crystal Palace events were cheek by jowl with announcements for the other entertainment venues in the city: Niblo's Garden, Castle Garden, and the theaters. The highlight of Barnum's tenure as head of the association was the Fourth of July festivities at the Crystal Palace. On the front page of the *New York Tribune*, Barnum announced that the holiday would be celebrated "in good old fashioned style." The day began with the Crystal Palace bell ringing for one hour and ended with evening fireworks accompanied by music.[106] But even Barnum's talents for entertainment were insufficient to keep the exhibition alive. Debts

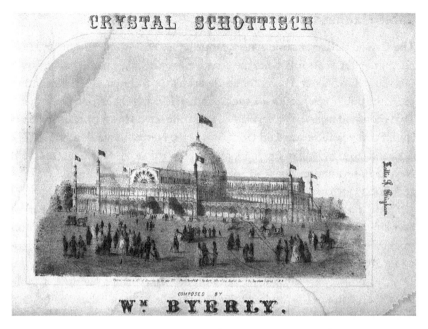

FIGURE 6.1. William Byerly, "Crystal Schottisch" (New York: W. A. Pond, 1853). Author's collection.

were heavy, and in its second year, many exhibitors did not return. Barnum resigned as head of the association at the end of July, just three months after the reopening. In November the exhibition closed for good.[107] As a spectacle, the Exhibition of the Industry of All Nations was a temporary attraction. But the Crystal Palace lived on.

As the largest enclosed space in the city, the Crystal Palace continued to be a useful venue until it burned down in October 1858. Promotion of the exhibition had ensured that the building was a tourist attraction in and of itself. The Crystal Palace was the largest building in the United States and an engineering achievement that showcased America's arrival as a technological and industrial competitor with Europe. Souvenirs from the exhibition put the iconic building onto everyday objects: wall clocks, sewing boxes, letter sheets, window shades and cotton handkerchiefs. Those who did not attend the exhibition (which was most of America) could still obtain a lasting image of the Crystal Palace on color lithographs. Maps of the city that highlighted attractions included the Crystal Palace. The sheet music for dances composed to celebrate the exhibition—the "Crystal Schottisch," the "Crystal Palace Cotillion," and the "Crystal Palace Polka"—were illustrated with an image of the building.[108]

Cable Madness

The Crystal Palace became synonymous with American invention and production. In the years following the closing of the exhibition, the building was the venue for fairs and ceremonies dedicated to technology. In September 1858, the public celebration of the Atlantic telegraph cable took place there. Two thousand miles of underwater cable connected Newfoundland and Ireland. Further cables was laid from St. John's, Newfoundland, to New York City, and from Ireland to England, making it possible to relay messages from New York to London. As proof of the cable's success, President Buchanan and Queen Victoria exchanged messages across the ocean.[109]

Cable madness, similar to the elation experienced seventy years earlier over air balloons, seized the nation. Just as Americans had purchased balloon stockings, hats, and accessories, one cartoonist suggested a "New and Appropriate Style of Costume—A LA CABLE!" for men's trousers.[110]

Sheet music, lavishly illustrated with scenes of ships laying the cable filled shop windows. Americans who had danced to the "Alida Waltz" and the "Crystal Schottisch" now dipped and whirled to the "Atlantic Telegraph Schottische," "Ocean Cable Polka," "Atlantic Telegraph Polka," and the "Ocean Cable Gallop."[111] Merchants capitalized on cable enthusiasm: pills, hats, and even insect powder grabbed the public's attention with the magic word "cable" in their advertisements. Brandreth's Pills simply used the screamer "Ocean Telegraph!" Knox hats informed *New York Daily Tribune* readers, "The Cable is Laid, the celebration in Commemoration of the

FIGURE 6.2. *Harper's Weekly*, September 4, 1858. Library Company of Philadelphia.

Figure 6.3. Atlantic cable section. Division of Work and Industry, National Museum of American History, Smithsonian Institution.

Event is Over, the Watering Places all Deserted, the City is Crowded, and Knox's Fall Style of Hat is Out." Lyon's Magnetic Powder and Pills for Insects and Vermin highlighted their advertisement with a poem, "The Cable Forever." Chemists carried the Atlantic Cable Bouquet perfume, "distilled from ocean spray and fragrant flowers."[112] *Harper's Weekly* published a "Great Telegraph" supplement.[113] Tiffany and Company sold souvenir pieces of the cable (lengths that had lain on the ocean floor before being replaced) for twenty-five cents each. *Knickerbocker Magazine* offered new subscribers a generous two feet of the cable with a certificate from Tiffany's confirming its authenticity.[114]

The Crystal Palace ceremony crowned a day of festivities and a grand procession from the Battery to Fortieth Street. Flags, banners, and transparencies embellished shop fronts along the procession route.[115] Tiffany's storefront was draped with cable, no doubt as a reminder that they held the exclusive right to sell it as souvenirs. Tributes to the progress of American science graced several buildings: a grand arch on E. Haughwout and Company's store held a stained-glass image of Benjamin Franklin flanked by Samuel F. B. Morse (inventor of the code used for telegraphs) and Cyrus Field (president of the American Telegraph Company, and the man responsible for laying the cable). Waterbury's furniture store displayed a banner with the following notice:

MARRIED,
On Thursday, August 5th, 1858, in the Church of Progress, at the Altar of Commerce, the Old with the New World. May they never be divorced.
BORN,
August 16th, 1858, in the Bed of the Ocean of Science and Enterprise, The Atlantic Telegraph. May it live to honor its parents.
DIED,
Monday, August 16th, 1858, from an Electric Shock, Old Fogyism. May he rest in peace.
We rejoice at the Marriage, the Birth, and the Death.[116]

Barnum's Museum exhibited a transparency "of the two steam frigates splicing the cable, surrounded with medallion portraits of Buchanan, Victoria, Morse, Franklin, Hudson and Field." Triumphal arches stood at the south, east, and west gates to Central Park. The eastern arch showed a telegraph line and printing press. The inscription read: "'The paths of Franklin led to a Field of enterprise.'" The banner on Douglas & Co.'s Mercantile Agency proclaimed: "'Honor to Cyrus W. Field! Welcome to the new agency for the promotion and protection of trade between Europe and America.'"[117]

The procession itself was a miniature re-creation of the Exhibition of the Industry of All Nations: manufacturers of tools and machinery demonstrated their wares on platform cars. The American Express Company car, drawn by ten horses, displayed a coiled section of the Atlantic cable along with a "Printing Telegraph Machine, with an operator at work." The American Telegraph Company's Hughes Printing Instrument operated as the car made its way along Broadway. The New York Typographical Society car carried "an old wooden press, such as Franklin wrought at and a new single cylinder press of Hoe's make, on which was being struck off various announcements which were distributed among the crowd." The ink manufacturer Thaddeus David and Company's car held a twelve-foot-high ink bottle. Wheeler and Wilson displayed "fair maidens busily using the sewing machines made by the company." Ward's thread factory demonstrated "the process of spinning and winding sewing thread." Safes, agricultural implements, a "Gas-Making and Cooking Machine," and a grand piano passed along Broadway as tangible reminders of American technology.[118]

The grand celebration wended its way north to the Crystal Palace for the centerpiece of the celebration honoring the cable and the man who devised it. The program included the New York Harmonic Society singing an ode, "The Cable," to the tune of the "Star-Spangled Banner": "Now intellect reigneth from the earth to the skies, / And science links nations, that war shall not sunder." The band played both "God Save the Queen" and "Hail Columbia." David D. Field's "Oration on History of Telegraph Company" recited the trials, failures, and ultimate triumph of the Atlantic cable. Another song, "All Hail! An Ode," concluded the ceremony:

> All hail across the main!
> Thought thrills our cable chain,
> Hear, nations, hear!
> Mind is victorious,
> Columbia's made glorious,
> While God watch'd over us; Hear, nations, hear![119]

On this nationally important occasion, the Crystal Palace played a prominent role. Already recognized as the preeminent venue for American industrial and manufacturing displays, the Atlantic cable celebration sealed the building's reputation as a temple of science and technology.[120]

The Atlantic cable ceremony was the largest, and the last, festivity in the Crystal Palace. Just a month later, the building went up in flames. In a matter of hours, the glass and steel structure was reduced to a pile of melted rubble. The American Institute fair had opened in the building three weeks earlier. As the institute gained prominence, its annual fairs grew larger. Ever in need of more space for exhibitors, the organizers naturally turned to the building dedicated to showcasing science and technology. The fire destroyed the entire contents of the building, including all the goods, machines, and models on display. One vendor, Franklin Harvey Biglow, who was in the building when the fire started, later described to his sister "the crackling of glass and the roaring of wind and flames as they rushed up through the roof and sides of the building." He regretted that he had not insured his goods, but he added, "I supposed that there was little, or no risk in such a building."[121] George Templeton Strong, who saw the building a few hours after the fire, described what was left of the walls looking like "a Brobdignagian aviary."[122] Currier and Ives memorialized the conflagration with a lithograph, *Burning of the New York Crystal Palace*.[123]

Within days, the public was offered "interesting souvenirs of all that remains of the finest building ever erected in America": pieces of fused glass from the building and charred items from the American Institute exhibition.[124]

The Crystal Palace was a tangible symbol of American ambitions and ideals. The building provided opportunities for the articulation of national identity: a United States poised to achieve global economic and industrial

FIGURE 6.4. Currier and Ives. *Burning of the New York Crystal Palace: on Tuesday Oct. 5th, 1858.* (New York: Currier and Ives, 1858). Library of Congress, Prints and Photographs Division.

dominance. As E. H. Chaplin told his listeners at the Fourth of July celebrations in 1854, the Crystal Palace was an appropriate place to celebrate
the nation's birthday: "For around us are the best achievements of civilization; around us are the true vehicles of American power—around us are the
prophecies of the coming time. . . . Our land is the granary of the world.
Our ships cut a perpetual wake around the globe. The wilderness of the
Pacific grows populous in a day; while Japan opens its sullen gates to amity
and commerce."[125] Luther Marsh, at the opening ceremonies of the American Institute fair at the Crystal Palace in 1855, suggested that the building
might become "the perpetual home of American genius. Hitherward tend
all the contributions of American invention, so that if anyone would know
what America is, and can become, let him bend his steps to this Temple of
Uses."[126] The Franklin Institute and the American Institute engaged in many
activities that promoted American science and technology: fairs, publications, educational programs, and lobbying fostered talent and celebrated
genius. At the same time, these activities sought to protect and promote
American products. But it was the Crystal Palace that captured the public's
imagination and remained, long after its demise, a touchstone for national
self-definition.

Conclusion
The First American Century

On May 10, 1876, the International Exhibition of Arts, Manufactures, and Products of the Soil and Mine opened in Philadelphia. The Centennial Exhibition, as it was commonly called, was a national celebration of progress intended to "make evident to the world the advancement of which a self-governed people is capable." Years of planning and fund-raising resulted in the largest fair ever staged in the United States. Two hundred and fifty buildings were constructed on 285 acres in Philadelphia's Fairmount Park. Over thirty thousand exhibitors displayed their machines, raw materials, products, foodstuffs, and art. Seventeen states and nine foreign nations had their own buildings. Guidebooks, official and otherwise, provided maps, timetables (of when the Pennsylvania Railroad trains stopped at the exhibition station), information, and suggestions of what to see. For 150 days, from May to November, nearly ten million people visited the fair.[1]

Though international in scope, the Centennial Exhibition was first and foremost an American fair. The *New York Times* told its readers, "However much we may find to astonish and instruct us in the triumphs of the rest of the world at Philadelphia, the rest of the world will be still more struck with ours."[2] Thanks to the fairs of the Franklin institute, the American Institute, and to the New York Crystal Palace exhibition, Americans knew how to put on a show. In the Agricultural Hall, foods and fibers were lavishly displayed. Outside the building, manufacturers demonstrated the efficient farm machinery that made the abundance of grains and vegetables possible. The aptly named Machinery Hall emphasized the rapid advancement of steam technology: the fourteen-acre exhibition space held fire engines, road rollers, pile drivers, blast furnace blowers, and cables, gears, shafts, and belts all driven by steam power. The vast building had room for engines large and small, including several locomotives made by a local manufacturer, the Baldwin Locomotive Works. Baldwin locomotives transported goods and people all over the United States.[3] Another product of American engineering, the Corliss engine, was the largest steam engine in the world. The centerpiece of Machinery Hall, it powered all the machinery in the building.

The United States Patent Office had its own building at the Centennial Exhibition. Five thousand patent models (only 3 percent of all patents in the United States in 1876) showcased American ingenuity. Inventions on display included Mrs. Potts's "Cold Handle, Double Pointed" smoothing irons and Martha J. Coston's Pyrotechnic Night Signal (a device considered so useful that the US government purchased the patent rights for $20,000). These were two of the more than eighty patents taken out by women. The presence of such inventions confirmed that women as well as men benefited from training and opportunities to improve American technology.[4]

FIGURE 7.1.   Stereoscope image of the Corliss engine. Centennial Photographic Company, 1876. Author's collection.

On the grounds of the exhibition, visitors admired the sculptures scattered along the paths and around a five-acre lake. They stopped for refreshments at any of eight eating places, including the Great American Restaurant, which boasted that "every Article connected with this restaurant is of American manufacture."[5] Visitors climbed to the torch of Frédéric-Auguste Bartholdi's incomplete work, *Liberty Enlightening the World*. Conceived as a gift from France to the United States for the centennial, the Statue of Liberty, as it became known, was not erected in New York Harbor until 1886. But the arm holding the torch was shipped from France in time to contribute to the celebration of one hundred years of American independence.

Just as at the New York Crystal Palace, visitors took away tangible memories of the Centennial Exhibition. Souvenir vendors sold handkerchiefs, puzzle blocks, sheet music, and fans. A machine stamped out souvenir coins embossed with the exhibition buildings. The guidebook office sold blank notebooks to record notable sights and experiences. Local bookshops offered lavishly illustrated keepsake books to remind visitors (or to inform readers who had not been to the exhibition) of the Centennial Exhibition's opulent variety of goods, objects, and art as well as its impressive buildings. At the Centennial Photographic Company building, visitors had their photographs taken for their entry passes, and purchased images of the Centennial Exhibition buildings, grounds, and displays.

The Centennial Exhibition was the culmination of many of the initiatives begun at the nation's founding. The results of a national endeavor to

FIGURE 7.2.   Colossal hand and torch, Bartholdi's statue of "Liberty." Robert N. Dennis Collection of Stereoscopic Views, the Miriam and Ira D. Wallach Division of Art, Prints, and Photographs, New York Public Library.

FIGURE 7.3.   Centennial Exhibition handkerchief. Library Company of Philadelphia.

develop natural resources, improve agriculture, and increase manufacturing were proudly displayed there. Silk production, begun as a colonial enterprise and continued in the early republic as a symbol of national resourcefulness, had become an American industry by the 1870s. Much of the raw silk was imported; domestic sericulture declined significantly after the 1840s, and silk manufacturing paled in comparison with other American textiles.[6] Nonetheless, silk remained an aspirational product, and one that was proudly displayed at the Centennial Exhibition. L. P. Brockett, spokesperson for the Silk Association of America, articulated the hope that, in the near future, American silk would "greatly surpass the products of European looms."[7]

There was abundant evidence of the nation's natural heritage. In Agricultural Hall, visitors marveled at "Antedeluvian Relics"—life-size casts of extinct mammals. Jefferson's desire to prove that North American fauna were more impressive than those in Europe reached fulfillment when the

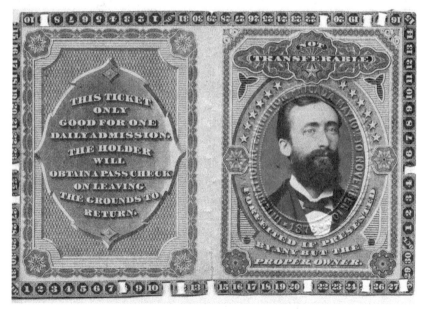

FIGURE 7.4. Thomas Janney's season ticket to the Centennial Exhibition. Author's collection.

FIGURE 7.5. Centennial souvenir printed on silk manufactured in the United States. Author's collection.

first remains of the megatherium (a giant ground sloth) were discovered in North America. Jefferson called it "great claw"; it was formally named *Megalonyx jeffersoni*. The bones of these twenty-foot-long creatures found in West Virginia and elsewhere joined the public displays of American mastodons.[8] Interest in American natural history quickened when the Academy of Natural Sciences in Philadelphia displayed *Hadrosaurus folkii* in 1869. This was the first dinosaur exhibited anywhere in the world. Excavated in Haddonfield, New Jersey, the extinct giant lizard drew unexpectedly large crowds to the academy. Almost one hundred thousand visitors (double the yearly number of visitors to the academy) came to see the specimen. *Hadrosaurus folkii* inaugurated "dinosauromania" in America.[9] The hadrosaur's popularity may have had something to do with media coverage in popular magazines such as *Scientific American* and *Harper's Monthly Magazine*. The *Harper's* article about the western expedition to uncover the bones of American monsters reads more like an adventure story than a scientific endeavor. Interspersed with an account of a titanotherium ("a monster of such vast proportions that a lower jaw measured over four feet in length") are descriptions of encounters with poisonous snakes and menacing bears.[10] Americans could view the titanotherium and other giants close to home; by the time the *Hadrosaurus* went on display in Philadelphia, colleges, museums, and private collectors had already acquired reproductions of extinct fauna. These were purchased from Ward's Natural Science Establishment, a business started by a professor of geology and zoology at Rochester University.[11] Centennial organizers, inspired by the *Hadrosaurus*'s popularity at the Academy of Natural Sciences and the public's interest in excavations in the far West, included a prehistoric exhibit in Agricultural Hall. Benjamin Hawkins, the designer of life-size models of extinct reptiles for the grounds of London's Crystal Palace, created a full-size *Hadrosaurus* for the Centennial Exhibition. Ward's company supplied a megatherium. For visitors who wanted a lasting memory, the Centennial Photographic Company sold stereograph images of this creature.[12] Even after the Centennial Exhibition concluded, dinosauromania inspired the bone hunter Edward Drinker Cope to commission a life-size model of an *Atlantosaurus* (first discovered in Colorado) for the Centennial Exhibition fairgrounds.[13] Like much else at the Centennial Exhibition, dinosaurs were large, impressive, and American.

Philadelphia's extensive water system ably supplied the needs of the Centennial Exhibition's restaurant kitchens, lavatories, hydraulic machinery, and ornamental fountains. On Fountain Avenue, one of the main thoroughfares through the exhibition grounds, the Centennial Fountain (sponsored by the Catholic Total Abstinence Union of America) had four

FIGURE 7.6. Stereoscope of the megatherium. Centennial Photographic Company, 1876. Author's collection.

drinking spouts. Another fountain, erected by the Grand Division of Sons of Temperance of the State of Pennsylvania, boasted "twenty-six self-acting spigots," supplying ice water.[14] All these hydraulic displays were fed by a waterworks built especially for the exhibition. A pumphouse erected on the bank of the Schuylkill River with "powerful pumping engines of the most approved pattern" supplied seven million gallons of water daily to the buildings and grounds.[15]

The Centennial Exhibition even had its own telegraph office, from where, thanks to the Atlantic cable, telegrams could be sent "to all parts of the world."[16] The newest communication technology, the telephone, had one of its earliest demonstrations at the exhibition. Another innovation, the steam-powered elevator, conveyed visitors to the top of the observation tower. Eliphalet Remington exhibited another technological wonder, his recently patented typewriter. For a small fee, visitors could have a message typed on this new mechanized writing machine. While visitors strolled the grounds and marveled at machinery, air balloons floated overhead. Although this aviation technology had not lived up to its initial promise, a different technology for flight, the airplane, was only a few decades in the future.

Phrenology was no longer in vogue in 1876, but evidence of internal improvements through public initiatives was exhibited by the Education Office in the United States Government building. Models of schoolhouses and college buildings, catalogs of schools, colleges, and benevolent institutions and textbooks "showing the progress made in these works from those used at the opening of the century to those used today," testified to the public and private efforts to cultivate American talent.[17] But racial hierarchies and prejudices were still firmly in place. An exhibit labeled "Ethnology of the United States" displayed artifacts, photographs, and wax figures of "the savage tribes of men [who] are rapidly disappearing from our continent."[18] Just as Black Hawk confronted the buildings and technologies of white civilization in the 1830s, in 1876 the artifacts of Native American cultures were displayed at the Centennial Exhibition as a contrast to "the powers and resources of the United States."[19] After viewing the exhibit, the editor of the *Atlantic Monthly*, William Dean Howells, wrote, "The red man, as he appears in effigy and in photograph in this collection, is a hideous demon, whose malign traits can hardly inspire any emotion other than abhorrence."[20] Forty years on, publicly expressed racism was alive and well.

Black Americans had no official presence at the Centennial Exhibition. Some were there as visitors, though barely tolerated.[21] Black men held unskilled jobs at the exhibition as janitors, waiters, and messengers. The Restaurant of the South reified long-held prejudices. It advertised that "an 'old Plantation Darky Band' will furnish the music, and illustrate Southern plantation scenes."[22] The only recognition that all Black Americans were now free citizens was a sculpture, *The Freed Slave*, in Memorial Hall. In some white observers, this emblem of freedom evoked revulsion; one visitor from Ohio described it as "a very ugly negro."[23] Despite such racist comments, a popular illustrated book of the Centennial Exhibition, *Frank Leslie's Historical Register of the United States Centennial Exposition, 1876*, included a scene of well-dressed African American men, women, and children admiring the sculpture of a triumphant ex-slave freed from his shackles. Nevertheless, the science of man still privileged white northern Europeans over all other cultures and races. And in the post–Civil War United States, a new era of discrimination, fear, and violence was validated by the scientific racism first promoted by phrenologists.

Many of the exhibitors, prize winners, and patent holders were the beneficiaries of the formal and informal training promoted by the mechanics organizations and other science and technology societies in the first half of the nineteenth century. By the 1860s, these educational initiatives had resulted in the creation of land grant colleges. Government funding supported instruction in "branches of learning as are related to agriculture and the mechanic arts, in such manner as the legislatures of the States may respectively prescribe, in order to promote the liberal and practical education of the industrial classes in the several pursuits and professions in

THE STATUE OF "THE FREED SLAVE" IN MEMORIAL HALL.

FIGURE 7.7. "The Statue of 'The Freed Slave' in Memorial Hall." *Frank Leslie's Historical Register of the United States Centennial Exposition, 1876* (New York: Frank Leslie's Publishing House, 1877). The Miriam and Ira D. Wallach Division of Art, Prints, and Photographs, New York Public Library.

life."[24] And the decades of effort on the part of the American Institute to promote domestic production and lobby for protective tariffs was finally rewarded: 1876 was the first year that the United States achieved a favorable balance of trade. Americans purchased more of their own products than foreign ones, and other nations, including Britain, bought significant amounts of the raw materials and machinery developed by Americans over the preceding century.[25] In 1853 American boosters at the Crystal Palace Exhibition heralded the rising economic power of the United States. A quarter century later, that somewhat premature boast was fact: the United States was an industrial and economic titan.[26] As the president of the Centennial Commission predicted in 1872, "Each State of the Union as a member of one united body politic, will show her sister States and to the world, how much she can add to the greatness of the nation of which she is a harmonious part."[27]

The results owed much to Americans' belief that developing technology was a national endeavor, a way to define what the country was and what its people were capable of. Americans wished to better themselves as individuals and as communities, and they wanted to compete in the world marketplace. Framing these desires as a national project spurred the activities, funding, and training necessary to make it happen. Thus, science and technology were front and center in public life: central to business, education, entertainment, and consumer culture.

We no longer purchase clocks embellished with air balloons, dinner plates with images of railway locomotives, or music celebrating water systems. The products that express national enthusiasm have changed: when the space race between the United States and the USSR made technology a weapon in Cold War politics in the mid-twentieth century, a rocket ship was the toy of choice for countless children. It was a tangible expression of the desire for American supremacy. In the twenty-first century, scholars and educators lament that Americans are falling behind in science- and technology-related skills and occupations. This woeful neglect is perceived as a national problem, making the United States less competitive in industry, agriculture, and medicine. With renewed local and national efforts to promote a greater investment in science and technology, we may again see the popular celebration of achievements—perhaps dining chairs emblazoned with iPhones or tableware sporting Teslas. A historical perspective on the public's involvement with science and technology informs our understanding of the modern embrace of innovation and how that enthusiasm continues to infuse American cultural practices and daily life.

ACKNOWLEDGMENTS

Research for this project was facilitated by several grants and fellowships. At Syracuse University, I am indebted to the Frank and Helen Pellicone Faculty Scholarship and the Appleby-Mosher Fund. Archive research was sponsored by the American Philosophical Society, the American Antiquarian Society's Kate B. and Hall J. Peterson Fund, and the Friends of the United States Air Force Academy Library's Clark-Yudkin Fellowship.

Archivists and librarians are very helpful people. The following individuals are especially so: Brianne Barrett, library and program assistant; Ashley Cataldo, curator of manuscripts; and Laura Wasowicz, curator of children's literature, all at the American Antiquarian Society; Courtney Christner at Independence National Historical Park; Beth Farwell, interim music liaison librarian, Crouch Fine Arts Library at Baylor University; Charles B. Greifenstein, associate librarian and curator of manuscripts at the American Philosophical Society; Lynn McCarthy at Winterthur Museum, Garden, and Library; and Mary Ruwell, chief of Special Collections, McDermott Library, at the United States Air Force Academy.

A good part of my research was carried out at the Library Company of Philadelphia. My thanks to the librarian James Green, the chief of reference Cornelia King, and the digitization and rights coordinator Emily Smith. And though he was unable to persuade me that weather should be a central theme in my study, Roy Goodman, former curator of printed materials at the American Philosophical Society, gave me much help and good advice.

My special thanks to Cynthia Kierner and Gregory Nobles, both of whom championed my efforts for more years than I care to admit. They encouraged me as this project was transformed from half-baked idea to publishable book. Michael McGandy, editor at Cornell University Press, and Clare Jones, assistant editor, made the publishing process a pleasure rather than a chore.

Many people and places contributed to my ability to put words on a page. My thanks to Alan Allport, Robyn Davis, Dan Gordon, Sally Gordon, the late C. Dallett Hemphill, Samantha Herrick, Susan Klepp, Morgan Kolakowski, Norman Kutcher, Leeyanne Moore, Simon Newman, Diana Norcross, Erin

Smith, and Michael Wilson for encouragement, constructive criticism, and friendship. Tom Humphrey must be singled out for thanks not only because he is a good historian but also because he keeps me entertained with his snarky wit and encyclopedic knowledge of professional cycling. Closer to home, my friends and neighbors Lori Brown and Martin Hogue supplied me with baked goods, whiskey, and good company. My father, George Branson, who is well into his tenth decade, helped me keep the goal of completing this task in my sight line.

Mark, Kaitlin, and Zoe—thank you for your patience and encouragement. And finally, to the memory of absent friends Stache and Jake.

NOTES

Introduction

1. John Murrin, "A Roof without Walls," in *Beyond Confederation: Origins of the Constitution and American National Identity*, ed. Richard R. Beeman, Stephen Botein, and Edward C. Carter II (Williamsburg, VA: Omohundro Institute of Early American History and Culture, 1987), 344.

2. Benjamin E. Park, *American Nationalisms: Imagining Union in the Age of Revolutions, 1783–1833* (New York: Cambridge University Press, 2018), 9. Government policies, most notably the patent office and state-sponsored initiatives, were directed at invention and improvement.

3. Neil McKendrick, *The Birth of a Consumer Society: The Commercialization of Eighteenth-Century England* (Bloomington: Indiana University Press, 1982), 9.

4. Bernard Bailyn, "1776: A Year of Challenge—a World Transformed," *Journal of Law and Economics* 19, no. 3 (1976): 447.

5. T. H. Breen, "An Empire of Goods: The Anglicization of Colonial America, 1690–1776," *Journal of British Studies* 25, no. 4 (October 1986): 493.

6. David Jaffe, "The Ebenezers Devotion: Pre- and Post-Revolutionary Consumption in Rural New England," *New England Quarterly* 76, no. 2 (June 2003): 254.

7. Appleby describes these artisan-capitalists as part of a "perpetually improving social engine" in which "farmers, lawyers, schoolteachers, manufacturers, and merchants—interacted to create general prosperity." Joyce Oldham Appleby, *Inheriting the Revolution: The First Generation of Americans* (Cambridge, MA: Harvard University Press, 2000), 255.

8. Mary Ann James, "Engineering an Environment for Change: Bigelow, Pierce and Early Practical Education at Harvard," in *Science at Harvard University: Historical Perspectives*, ed. Clark A. Elliott and Margaret W. Rossiter (Bethlehem, PA: Lehigh University Press, 1992), 59.

9. James A. Delbourgo, *A Most Amazing Scene of Wonders: Electricity and Enlightenment in Early America* (Cambridge, MA: Harvard University Press, 2007); Susan S. Parrish, *American Curiosity: Cultures of Natural History in the Colonial British Atlantic World* (Chapel Hill: University of North Carolina Press, 2006); Londa L. Schiebinger and Claudia Swan, eds., *Colonial Botany: Science, Commerce, and Politics in the Early Modern World* (Philadelphia: University of Pennsylvania Press, 2005).

10. Simon Schaffer, "Natural Philosophy and Public Spectacle in the Eighteenth Century," *History of Science* 21, no. 1 (1983): 1–43; Fred Nadis, *Wonder Shows: Performing Science, Magic, and Religion in America* (New Brunswick, NJ: Rutgers University Press, 2005); Ralph O'Connor, *The Earth on Show: Fossils and the Poetics of Popular Science, 1802–1856* (Chicago: University of Chicago Press, 2007); Paul Semonin, *American*

Monster: How the Nation's First Prehistoric Creature Became a Symbol of National Identity (New York: NYU Press, 2000).

11. Timothy Dwight, *Travels in New-England and New York*, vol. 1 (New Haven, CT: Timothy Dwight, 1821), 16, quoted in Arthur Alphonse Ekirch Jr., *The Idea of Progress in America, 1815–1860* (New York: AMS, 1969), 41.

12. Kariann Akemi Yokota, *Unbecoming British: How Revolutionary America Became a Postcolonial Nation* (Oxford: Oxford University Press, 2011).

1. Domestic Science

1. Eighteenth- and early nineteenth-century Americans did not use the term *science*. Instead, the term *natural philosophy* encompassed the knowledge, investigation, and observation of the natural world, from chemistry to geology, astronomy, physics, and botany.

2. Appleby, *Inheriting the Revolution*, 249.

3. William Penn (1682), quoted in Meyer Reinhold, "The Quest for 'Useful Knowledge' in Eighteenth-Century America," *Proceedings of the American Philosophical Society* 119, no. 2 (April 16, 1975): 110.

4. Reinhold, "Quest for 'Useful Knowledge,'" 114–15.

5. John H. Pollack, introduction to *"The Good Education of Youth": Worlds of Learning in the Age of Franklin*, ed. John H. Pollack (New Castle, DE: University of Pennsylvania Libraries and Oak Knoll Press, 2009), 4.

6. Reinhold, "Quest for 'Useful Knowledge,'" 117.

7. Martha Logan and Mary Barbot Prior, "Letters of Martha Logan to John Bartram, 1760–1763," *South Carolina Historical Magazine* 59, no. 1 (January 1958): 38–46; Susan Branson, "Flora and Femininity: Gender and Botany in Early America," *Commonplace* 12, no. 2 (January 2012), http://commonplace.online/article/flora-femininity/.

8. Raymond Phineas Stearns, *Science in the British Colonies of America* (Urbana: University of Illinois Press, 1970), 559–618; Branson, "Flora and Femininity"; Parrish, *American Curiosity*, 196–97; Sara Stidstone Gronim, "What Jane Knew: A Woman Botanist in the Eighteenth Century," *Journal of Women's History* 19, no. 3 (Fall 2007): 33–59.

9. Susanna Wright, "Directions for the Management of Silk Worms. By the Late Mrs. S. Wright, of Lancaster-County, in Pennsylvania," *Philadelphia Medical and Physical Journal* 1 (1804): 103–7. Benjamin Smith Barton's introduction to Wright's posthumously published essay praises her as "a woman of uncommon powers of mind" who "directed much of her attention to the management of silk-worms and to other subjects of public utility at a time (at least forty years ago) when she stood alone in her exertions in this way" (103).

10. David L. Coon, "Eliza Lucas Pinckney and the Reintroduction of Indigo Culture in South Carolina," *Journal of Southern History* 42, no. 1 (February 1976): 61–76.

11. Zara Anishanslin, *Portrait of a Woman in Silk: Hidden Histories of the British Atlantic World* (New Haven, CT: Yale University Press, 2016), 310; Ben Marsh, "Silk Hopes in Colonial South Carolina," *Journal of Southern History* 78, no. 4 (November 2012): 847.

12. Daniel Leeds, *The American Almanack for the Year of Christian Account, 1712* (New York: William and Andrew Bradford, 1712), quoted in Sara Stidstone Gronim, *Everyday Nature: Knowledge of the Natural World in Colonial New York* (New Brunswick, NJ: Rutgers University Press, 2007), 97.

13. Robyn Davis McMillin, "Science in the American Style, 1700–1800" (PhD diss., University of Oklahoma, 2009), 108.

14. David A. Copeland, *Colonial American Newspapers: Character and Content* (Newark: University of Delaware Press, 1997), 279; Richard R. Johns, *Spreading the News: The American Postal System from Franklin to Morse* (Cambridge, MA: Harvard University Press, 1998), 24. Newspapers reflected the everyday world in addition to political essays and legislative reports. Also, "newspapers narrowed the gap between the learned and the merely literate and the information gap between the privileged and the merely competent." Charles E. Clark, "The Newspapers of Provincial America," in *Three Hundred Years of the American Newspaper: Essays*, ed. John B. Hench (Worcester, MA: American Antiquarian Society, 1991), 387, 385.

15. In reality, Keimer simply copied articles from *Chambers Dictionary*. C. Lennart Carlson, "Samuel Keimer: A Study in the Transit of English Culture to Colonial Pennsylvania," *Pennsylvania Magazine of History and Biography* 61, no. 4 (1937): 381.

16. James Raven, *London Booksellers and American Customers: Transatlantic Literary Community and the Charleston Library Society, 1748–1811* (Columbia: University of South Carolina Press, 2002), 62; James Raven, "Social Libraries and Library Societies in Eighteenth-Century North America," in *Institutions of Reading: The Social Life of Libraries in the United States*, ed. Thomas Augst and Kenneth Carpenter (Amherst: University of Massachusetts Press, 2007), 34.

17. David Kaser, *A Book for a Sixpence: The Circulating Library in America* (Pittsburgh: Beta Phi Mu, 1980), 117.

18. Thomas Paine, "Philosophical Queries," *Pennsylvania Magazine, or, American Monthly Museum* 1 (August 1, 1775): 353.

19. Frank Luther Mott claims that the articles by prominent Philadelphia scientists "helped to give the Pennsylvania Magazine a notably scientific trend." Mott, *A History of American Magazines*, vol. 1, *1741–1850* (Cambridge, MA: Belknap Press of Harvard University Press, 1930), 90.

20. "A Description of a New Invention of a Spinning Machine," *Pennsylvania Magazine, or, American Monthly Museum* 1 (April 1, 1775): 158. The Declaration of Independence appeared in the final issue in July 1776. Neil Longley York says Paine and other contributors "wanted their readers to develop a national identity, and promoting invention was one way of doing it." Neil Longley York, *Mechanical Metamorphosis: Technological Change in Revolutionary America* (Westport, CT: Greenwood, 1985), 55.

21. Elizabeth Carroll Reilly and David D. Hall, "Customers and the Market for Books," in *The Colonial Book in the Atlantic World*, ed. David Hall and Hugh Amory (Chapel Hill: University of North Carolina Press in association with the American Antiquarian Society, 2007), 392.

22. McMillin, "Science in the American Style," 116.

23. Molly A. McCarthy, "Redeeming the Almanac: Learning to Appreciate the iPhone of Early America," *Commonplace* 11, no. 1 (October 2010), http://commonplace.online/article/redeeming-the-almanac/; Molly A. McCarthy, *The Accidental*

Diarist: A History of the Daily Planner in America (Chicago: University of Chicago Press, 2013).

24. Silvio Bedini, *Benjamin Banneker* (Rancho Cordova, CA: Landmark Enterprises, 1972), 136.

25. Bedini, 52.

26. *Benjamin Banneker's Pennsylvania, Delaware, Maryland and Virginia Almanac and Ephemeris for the Year of Our Lord, 1792* (Baltimore: William Goddard and James Angell, 1791), 2.

27. Charles Cerami, *Benjamin Banneker: Surveyor, Astronomer, Publisher, Patriot* (New York: John Wiley & Sons, 2002), 150.

28. Thomas Jefferson to Benjamin Banneker, August 30, 1791, *The Papers of Thomas Jefferson Digital Edition*, ed. Barbara B. Oberg and J. Jefferson Looney (Charlottesville: University of Virginia Press, Rotunda, 2008), accessed September 23, 2019, http://rotunda.upress.virginia.edu:8080/founders/default.xqy?keys=TSJN-print-01-22-02-0091 (original source: main series, vol. 22 [August 6–December 31, 1791]).

29. Cerami, *Benjamin Banneker*, 173.

30. Further pages on lightning and electricity briefly mentioned Franklin's experiments, but most of the article was devoted to an account of a Mr. de Romas's 1753 kite experiment.

31. Benjamin Franklin, *Poor Richard Improved; Being an Almanac [. . .] for the Year of Our Lord 1753* (Philadelphia: Franklin and Hall, 1752).

32. Franklin, *Poor Richard Improved*. Robyn Davis McMillin notes that by midcentury almanac makers provided an education in astronomy to readers, thus "attesting to the popular interest in science as well as the extensive instruction in science available through unconventional or alternative means." "Science in the American Style, 1700–1800," 124.

33. "To the Printers," *Boston Gazette, or Weekly Journal*, February 7, 1757.

34. Rebecca Yamin, *Digging in the City of Brotherly Love* (New Haven, CT: Yale University Press, 2008); Patrice L. Jeppson, "Comets and Calendars," talk presented as part of Explore Philadelphia's Hidden Past: A Pennsylvania Archaeology Month Celebration, sponsored by the Philadelphia Archaeological Forum and Independence National Historical Park, Philadelphia, 2007; Silke Ackerman, "Maths and Memory: Calendar Medals in the British Museum, Part 1," *Medal* 45 (2004): 3–44.

35. Benjamin Martin, *Supplement Containing Remarks on a Rhapsody of Adventures of a Modern Knight-errant* (Bath: printed for the author, 1746), 28–29, quoted in Simon Schaffer, "Natural Philosophy and Public Spectacle in the Eighteenth Century," *History of Science* 21 (1983): 10. Increase Mather's *Kometographia, or, A discourse concerning comets [. . .]* (Boston: Samuel Green, 1683) is an example of the pre-Enlightenment signs and omens interpretation of celestial events.

36. McMillin, "Science in the American Style," 121. J. Rixby Ruffin describes this transition as "a hybrid cosmology that, while gradually introducing new scientific thought, neither contradicted their secular knowledge nor offended their spiritual beliefs." Ruffin, "'Urania's Dusky Vails': Heliocentrism in Colonial Almanacs, 1700–1735," *New England Quarterly* 70, no. 2 (June 1997): 312.

37. Richard Draper, *Blazing-stars messengers of God's wrath: in a few serious and solemn meditations upon the wonderful comet [. . .].* (Boston: printed and sold by R. Draper in Newbury-Street; and by Fowle and Draper in Marlborough-Street, 1759).

38. The *Pennsylvania Gazette* published a detailed description of an orrery and its uses in 1771. "Extract from the Journals," *Pennsylvania Gazette* (Philadelphia), March 28, 1771.

39. Not all owners were men. In Philadelphia, transit of Venus observers borrowed a telescope that belonged to Mary Norris. Benjamin Franklin purchased the refracting telescope for Isaac Norris, who bequeathed it to his daughter. American Philosophical Society, *Transactions of the American Philosophical Society, Held at Philadelphia, for Promoting Useful Knowledge* (1771), 1:44.

40. Philip Thomas Lee, Anne Arundel County Maryland, Charles County Maryland Register of Wills 1785–1791, 246–60, ordered July 20, 1779, taken July 30, 1780, recorded December 8, 1789; George Johnston, Fairfax County, Virginia (Alexandria), Fairfax County Will Book C-1 (1767–1776), 1–6, taken February 11, 1767, entered May 20, 1767. Probing the Past: Virginia and Maryland Probate Inventories, 1740–1810, http://chnm.gmu.edu/probateinventory/index.php.

41. *New-York Gazette*, March 18, 1765.

42. "To the Author of the New-England Courant," *New-England Courant* (Boston), December 3, 1722.

43. McMillin, "Science in the American Style," chap. 3.

44. William Smith, Provost of the College of Philadelphia, "An Account of the Transit of Venus over the Sun's Disk, as observed at Norriton, in the County of Philadelphia, and Province of Pennsylvania, June 3rd, 1769," in American Philosophical Society, *Transactions* (1771), 1:24.

45. I borrow the term *retrospective simultaneity* from Robyn McMillin. See "Science in the American Style," chaps. 3 and 5.

46. Roger Sherman, *An Astronomical Diary* (New York: Henry De Foreest, 1752).

47. Job Shepherd, *Poor Job, 1753* (Newport, RI: James Franklin, 1752).

48. Franklin, *Poor Richard Improved* [. . .] *1753*.

49. *Pennsylvania Gazette* (Philadelphia), September 6, 1753.

50. "Eclipses for 1753," *Poor Richard Improved* [. . .] *1753*.

51. "London," *Newport (RI) Mercury*, June 25, 1764.

52. "The Royal Society in London, and the Foreign Professors in Astronomy, request the Curious in North America, on the Continent, to observe the Transit of mercury over the Sun, which will happen on the 6th of May next." *Pennsylvania Gazette* (Philadelphia), April 5, 1753.

53. Brooke Hindle, *The Pursuit of Science in Revolutionary America, 1735–1789* (Chapel Hill: University of North Carolina Press, 1956), 147; "Royal Society," *Pennsylvania Gazette* (Philadelphia), April 5, 1753; Andrea Wulf, *Chasing Venus: The Race to Measure the Heavens* (New York: Knopf, 2012), 129. The last transit of Venus in the twenty-first century occurred in June 2012. The next transit will take place in 2117.

54. Sylvanus Urban, "History of the Late Comet," *Gentleman's Magazine* (London), November 1, 1759, 521–24.

55. *New-York Gazette*, December 10, 1764.

56. Gronim, *Everyday Nature*, 99; *New-York Mercury*, December 24 and 31, 1764.

57. The American Philosophical Society was organized in 1743 but soon languished. It was revived in 1767 at the same time a similar organization, the American Society, came into being. The two organizations merged in December 1768 to become

the American Philosophical Society, held at Philadelphia, for Promoting Useful Knowledge. See Hindle, *Pursuit of Science in Revolutionary America*, 127–36.

58. American Philosophical Society, *Transactions* (2nd ed., 1789), 1:xvii, xxi, xxii. The earlier American Society's minutes recorded examining domestic hemp fabric "equal in goodness to any of the kind imported into England." Murphy D. Smith, *A Museum: The History of the Cabinet of Curiosities of the American Philosophical Society* (Philadelphia: American Philosophical Society, 1996), 139. Quote is from the American Society *Minutes*, 1768. For the connection between useful knowledge and the eighteenth-century concept of polite culture as it was expressed by members of the APS, see Elizabeth E. Webster, "American Science and the Pursuit of 'Useful Knowledge' in the Polite Eighteenth Century, 1750–1806," PhD diss., University of Notre Dame, 2010.

59. American Philosophical Society, *Transactions* (1771), 1:xviii, xx. The first volume contained an essay from a New Jersey APS member, Edward Antill, on vine cultivation and wine production suited to North America. "An Essay on the Cultivation of the Vine, and the making and preserving of Wine, suited to the different Climates in North-America," *Transactions* (1771), 1:180–262.

60. Moses Bartram experimented with sericulture in the 1760s and 1770s. His essay, "Observations on the Native Silk-Worms of North America," was published in American Philosophical Society, *Transactions* (1771), 1:294–301; Anishanslin, *Portrait of a Woman in Silk*, 301; Ben Marsh, *Unravelled Dreams: Silk and the Atlantic World, 1500–1840* (Cambridge: Cambridge University Press, 2020).

61. Marsh, "Silk Hopes in Colonial South Carolina," 833, 839.

62. Anishanslin, *Portrait of a Woman in Silk*, 304.

63. American Philosophical Society, *Transactions* (1771), 1:xxiv. In 1786 the APS gave further encouragement to American mechanics with the offer of a two-hundred-guinea prize for the "best discovery, or most useful invention." "Account and Conditions of a Premium Offered by the American Philosophical Society," *Columbian Magazine* 1, no. 4 (December 1786): 179.

64. Hindle, *Pursuit of Science in Revolutionary America*, 193.

65. *A Catalogue of Instruments and Models in the Possession of the American Philosophical Society*, compiled by Robert P. Multhauf, assisted by David Davies (Philadelphia: American Philosophical Society, 1961), 40, 44–46.

66. The South Carolinians William Byrd II and Robert Carter both purchased Elizabeth Carter's translation of Newton for their daughters. Parrish, *American Curiosity*, 184. There were advertisements for the book in the *South Carolina Gazette* in the 1750s.

67. Barbara Maria Stafford, *Artful Science: Enlightenment Entertainment and the Eclipse of Visual Education* (Cambridge, MA: MIT Press, 1994), 58.

68. Geoffrey Sutton, *Science for a Polite Society: Gender, Culture, and the Demonstration of Enlightenment* (New York: HarperCollins, 1995), 227–28.

69. Bedini, *Benjamin Banneker*, 73–74.

70. Joseph Priestley, *A Familiar Introduction to the Study of Electricity* (London: J. Johnson, 1768), 10. The presence of learned women in intellectual life did not go unnoticed. George Ballard's *Memoirs of British Ladies Who Have Been Celebrated for Their Writings or Skill in the Learned Languages, Arts and Sciences* (1775) surveys women's contributions to the sciences from the fourteenth to the eighteenth century.

Thomas Amory's picaresque tale, *The Life of John Buncle, Esq; Containing Various Observations and Reflections, made in Several Parts of the World, and Many Extraordinary Relations* (1770), includes an account of a utopian community of one hundred women who devote themselves to the study of natural philosophy. George (or Martha) Washington owned Amory's book. George Washington, Fairfax County, Virginia, Will Book J 1 1801–1806, fol. 326, death date December 14, 1799, entered August 20, 1810. Probing the Past: Virginia and Maryland Probate Inventories, 1740–1810, http://chnm.gmu.edu/probateinventory/index.php. But the woman of science also became a stock character in French and English plays, beginning in the 1670s with Molière's *Les femmes savantes* (1672) and Thomas Wright's *The Female Virtuosos* (1693). These portrayals treated learned women as objects of ridicule rather than legitimate participants in the project of enlightenment. James Miller's *The Humours of Oxford* (1730), for example, contains an unsympathetic portrayal of a middle-aged matron who, through scientific dilettantism, becomes a nuisance to society. Even female playwrights chose to make these women figures of fun: the heroine of Susannah Centlivre's comedy *The Basset-Table* (1705) loves her microscope more than her suitor. The American author Judith Sargent Murray, otherwise known for her championing of women, continued this tradition in her play *The Traveler Returned* (1796). Susan Scott Parrish links these unflattering, satirical portrayals of learned women to the tradition of the (often fatally) curious women in conduct books and fairy tales (especially Bluebeard). Parrish, *American Curiosity*, 175, 183. Novelists also criticized learned women: Jane Austen's character Mary Bennett in *Pride and Prejudice*, for example, is a pedantic nuisance to her family. See Jennifer Van Horn, *The Power of Objects in Eighteenth-Century British America* (Chapel Hill: published for the Omohundro Institute of Early American History and Culture, Williamsburg, Virginia, by the University of North Carolina Press, 2017).

71. Karin A. Wulf, introduction to *Milcah Martha Moore's Book: A Commonplace Book from Revolutionary America*, ed. Catherine La Courreye Blecki and Karin A. Wulf (University Park: Pennsylvania State University Press, 1997), 29.

72. "The prophetick Muse: To David Rittenhouse Esqr. By the same. May 1776," in Blecki and Wulf, *Milcah Martha Moore's Book*, 257–58.

73. "Copy of a Letter from M[.] Morris" (dated 1755), in Blecki and Wulf, *Milcah Martha Moore's Book*, 227.

74. "Copy of a Letter from M[.] Morris," 227–28.

75. "Copy of a Letter from M[.] Morris," 228.

76. Kaser, *Book for a Sixpence*, 117, 34. The catalog for the Library Company of Philadelphia contains evidence that women were readers there, even if they were not members. For example, in 1741 the library owned the following: *The Ladies Library; containing excellent Discourses upon several Vices and Virtues relating to the Sex; and Instructions for their Conduct, applied to several States, of the Daughter, the Wife, the Mother, the Widow, the Mistress &c. Written by a Lady* (London, 1732); and George Hicks, *Instructions for the Education of a Daughter* (London, 1721): see *Catalogue of the Library Company of Philadelphia* (1741). The Library Company's Shareholder's Book has entries for women beginning in 1769.

77. Nollet's *Leçons de physique expérimentale* is listed in the Library Company of Philadelphia catalog for 1770.

78. Benjamin Rush, *Syllabus of lectures containing the application of the principles of natural philosophy, and chemistry, to domestic and culinary purposes: Composed for the use of the Young Ladies' Academy, in Philadelphia* (Philadelphia: printed for Andrew Brown, 1787). Rush also believed that women should have "a general acquaintance with the first principles of astronomy, and natural philosophy." Rush, "Thoughts upon Female Education, accommodated to the Present State of Society, manners, and Government, in the United States of America [. . .]. (Philadelphia, 1791).

79. Rush, *Syllabus*. At Mrs. Bazeley's Seminary for Young Ladies, for example, the curriculum included botany and chemistry, and "Lectures on Geography, Astronomy and Natural Philosophy," *Aurora General Advertiser* (Philadelphia), September 2, 1812. See Jessica C. Linker, "The Fruits of Their Labor: Women's Scientific Practice in Early America, 1750–1860" (PhD diss., University of Connecticut, 2017).

80. Tucker's lectures were advertised in the *Aurora* on October 24, 1811. The following year, a Dr. Jones advertised chemical lectures. The price of a ticket was "for a gentleman $12, For a lady $6," *Aurora General Advertiser* (Philadelphia), November 12, 1812.

81. *A Grammar of Chemistry Wherein the Principles of the Science Are Familiarized by a Variety of Easy and Entertaining Experiments with Questions for Exercise, and a Glossary of Terms in Common Use by D. Blair Corrected and Revised by Benjamin Tucker* (Philadelphia: David Hogan, 1810). Some of these experiments may have been rather hazardous: Samuel Parkes's *A Chemical Catechism for the Use of Young People* (London: printed for the author, 1806) included a chapter titled "Amusing Experiments." Included was the following: "Take ten grains of oxygenized muriate of potass[ium?] and one grain of phosphorus. . . . [Rub them together in a mortar] . . . and very VIOLENT DETONATIONS will be produced. *In this experiment it would be dangerous to employ a larger quantity of phosphorus than that prescribed*" (542).

82. *The Diary of Elizabeth Drinker*, ed. Elaine Forman Crane, vol. 3 (Boston: Northeastern University Press, 1991), 1526, 1595.

83. Rachel Van Dyke, *To Read My Heart: The Journal of Rachel Van Dyke, 1810–1811*, ed. Lucia McMahon and Deborah Schriver (Philadelphia: University of Pennsylvania Press, 2000), 47. Augustus Van Dyke was practicing medicine in Philadelphia. He had studied under Benjamin Rush.

84. Van Dyke, *To Read My Heart*, 63, entry for June 29, 1810.

85. Van Dyke, 88, entry for July 25, 1810.

86. Esther Edwards Burr to Sarah Prince, June 14, 1755, in *The Journal of Esther Edwards Burr, 1754–1757*, ed. Carol F. Karlsen and Laurie Crumpacker (New Haven, CT: Yale University Press, 1984), 123.

87. Drinker, *Diary*, 3. Entries for February 8, 1760, and March 24, 1773. "Widow Bringhurst" was Mary Claypoole Bringhurst. Her son James (1730–1810?) was a contemporary of Elizabeth Sandwith. Bringhurst may have owned a static electricity generator, known as Priestley's machine, with a glass tube manufactured by Caspar Wister in the Philadelphia area. Delbourgo, *Most Amazing Scene of Wonders*, 130.

88. *Memoirs of the Forty-Five Years of the Life of James Lackington* (London: printed for the author, 1827), 236, quoted in Raven, *London Booksellers*, 173.

89. In a letter to his wife, Deborah, in 1758 Franklin listed items he had shipped to her. These included "a Mahogany and a little Shagrin Box with Microscopes and

other Optical Instruments loose, [which] are for Mr. Allinson if he likes them; if not, put them in my Room 'till I return." In 1770 Franklin ordered a reflecting telescope for Humphrey Marshall of Chester County, Pennsylvania. Franklin told Marshall that "Dr. Fothergill had since desired me to add a Microscope and Thermometer, and will pay for the whole." Benjamin Franklin, *The Papers of Benjamin Franklin* (New Haven, CT: Yale University Press, 1959), 7:379a, 17:109a. Transatlantic transportation was not without hazards. Peter Collinson informed Franklin that "Poor John Bartram has lost his Two Guinea Microscope" in a ship that sank in the Atlantic on its way from England to Philadelphia. Collinson to Franklin, January 14, 1753 [1754?], London, *Papers*, 5:188a.

90. *Pennsylvania Gazette* (Philadelphia), February 3, 1742. A notice of instruments for sale by Andrew Porter (in Union Street between Second and Third) included a solar telescope, solar microscope, thermometer, hygrometer, barometer, and manual orrery. *Pennsylvania Gazette*, July 13, 1774. Gardiner Baker advertised "Several Electrical machines, with Insulating Stock" for sale at his New York museum. *Diary or Loudon's Register* (New York), July 14, 1797.

91. *Georgia Gazette* (Savannah), December 1, 1763. Numerous advertisements in the *Pennsylvania Gazette* testify to curiosity about this instrument. For example, in August 1744, "The Solar or Camera Obscura MICROSCOPE . . . Just arrived from LONDON," could be viewed for eighteen pence by "Gentlemen and Ladies" of Philadelphia. Benjamin Franklin also printed a broadside to advertise the microscope. "Just arrived from London, for the entertainment of the curious and others, and is now to be seen, by six or more, in a large commodious room, at the house of Mr. Vidal, in Second-Street; the solar or camera obscura microscope" (Philadelphia, 1744), Early American Imprints, series 1, no. 5419.

92. Benjamin Franklin to John Winthrop, July 2, 1768, *Papers*, 15:166a. Though few colonists could afford these instruments, many were probably aware of their value (and resale value). A notice in the *Pennsylvania Gazette* for a stolen telescope indicates an understanding of these instruments as commodities: "3 pounds reward for stolen telescope.—stolen out of the house of William Wells, at the Old Ferry, in Water street, between Market and Arch streets." *Pennsylvania Gazette* (Philadelphia), November 1, 1770. Scientific instruments represented culture and status as well as knowledge: the popularization of the microscope and magnifying glass generated a vogue in miniature magnifiers worn on women's wrists. Parrish, *American Curiosity*, 59.

93. Elias Nason, *Sir Charles Henry Frankland, Baronet; or, Boston in the Colonial Times* (Albany, NY: J. Munsell, 1865), 81. Cited in John C. Autin, *Chelsea Porcelain at Williamsburg* (Williamsburg, VA: Colonial Williamsburg Foundation, 1977), 12.

94. Nicholas Flood, doctor, Richmond County, Virginia, Will Book #7, 1767–1787, 239–70, taken May 27, 28, 31 and June 1, 1776. Probing the Past: Virginia and Maryland Probate Inventories, 1740–1810, http://chnm.gmu.edu/probateinventory/index.php.

95. Joseph Pemberton, Anne Arundel County, Maryland, Maryland State Archives Testimentary Papers, box 4, folder 89, inventory entered December 11, 1783. Pemberton also owned a telescope. Peter Wagener, Fairfax County, Virginia, Will Book H 1, 140–44, ordered October 1798, taken January 9 and 10, 1799, recorded September 15, 1800. Probing the Past: Virginia and Maryland Probate Inventories, 1740–1810, http://chnm.gmu.edu/probateinventory/index.php.

96. Dr. John Stewart, Bladensburg, Prince George's County, Maryland, Prince George's Inventories 1796–1800, 186–89 (1797). Probing the Past: Virginia and Maryland Probate Inventories, 1740–1810, http://chnm.gmu.edu/probateinventory/index.php.

97. George Johnston, Fairfax County, Virginia (Alexandria), Fairfax County Will Book C-1 (1767–1776), 1–6; John West Jr. Fairfax County, Virginia, Will Book D 1, 123–27; Hannah Washington, Fairfax County, Virginia, Will Book J 1, 1806–1812, 19–27. Washington's "Dr. Priestley's Machine with Apparatus damage" was valued at £1. Probing the Past: Virginia and Maryland Probate Inventories, 1740–1810, http://chnm.gmu.edu/probateinventory/index.php. Thistlewood's instruments and Samuel Hayward's electrical machine are mentioned in Trevor Burnard, *Mastery, Tyranny, and Desire: Thomas Thistlewood and His Slaves in the Anglo-Jamaican World* (Chapel Hill: University of North Carolina Press, 2004), 119. Priestley's *History and Present State of Electricity* (1767) was the first comprehensive study of electricity. He published the less technical *A Familiar Introduction to the Study of Electricity* (1768) a year later for a more general readership. Priestley advertised his electrical machines at the back of *A Familiar Introduction to the Study of Electricity*. They were sold by J. Johnson at his bookstore near St. Paul's, London.

98. Bedini, *Benjamin Banneker*, 74.

99. "A Neat Pocket Microscope, with the Apparatus, & a fine Collection of Curiosities belong to the same," *Boston Evening-Post*, December 14, 1761; pocket microscope offered for sale in the *Royal Gazette* (New York), September 25, 1779; solar and pocket microscope advertised for sale in the *Independent Ledger and the American Advertiser* (Boston), September 4, 1786; "An elegant Planetarium, or small Orrery," *Catalogue of Mezzotinto and Copperplate Prints, Plate, and Pated Ware, &c. To be sold by Public Auction . . . At the late dwelling of Mrs. Pine* (Philadelphia, December 20, 1792); Edmund Smith advertised "one Electric Machine" along with "one good family sleigh" and "one second hand carriage," *Connecticut Herald* (New Haven), December 18, 1810.

100. "For the Telescope," *Telescope* (Columbia, SC), May 14, 1816.

101. Entry for May 14, 1765, "Extracts from the Journal of Mrs. Ann Manigault 1754–1781," *South Carolina Historical and Genealogical Magazine* 20 (1919): 208. Cited in Raven, *London Booksellers*, 174. Advertisement for William Johnson's demonstrations in *South Carolina Gazette* (Charleston), April 13, 1765.

102. "A Course of Natural Philosophy," *New York Gazette, or, the Weekly Post-Boy*, July 22, 1751.

103. Ebenezer Kinnersley's advertisements stated "to admit a gentleman and a lady, seven shillings and six pence; a single person five shillings."

104. Greenwood broadside, "Sublime Entertainment," Providence, Rhode Island, 1793, Early American Imprints, series 1, no. 46767.

105. There were other instances of enslaved men and women, perhaps unwillingly, experiencing the products of scientific enlightenment. In another part of the British Empire, the Jamaican planter Thomas Thistlewood recorded in his diary for March 30, 1772, that he took three of his slaves with him to visit his fellow planter, Samuel Haywood. "Vene, phibbah & John, down to See the governor &c. Mr. hayward Electrify'd them. gave them dinner, &c." These individuals were very likely

the subjects of a scientific experiment rather than keen participants and observers. Diary of Thomas Thistlewood, entry for Monday, March 30, 1772, Monson 31/23, p. 54, Lincolnshire County Archives, Lincoln, UK. My thanks to Trevor Burnard for providing me with this reference.

106. Christopher Looby, "The Constitution of Nature: Taxonomy as Politics in Jefferson, Peale, and Bartram," *Early American Literature* 22, no. 3 (1987): 253.

107. Broadside, "The American Museum," 1791, Early American Imprints, series 1, no. 23619; Charles Coleman Sellers, *Mr. Peale's Museum: Charles Willson Peale and the First Popular Museum of National Science and Art* (New York: W. W. Norton, 1980), 12, 52.

108. Broadside, "Museum & Waxwork," 1793, Early American Imprints, series 1, no. 25908.

109. Charles Willson Peale, *Introduction to a Course of Lectures on Natural History* (Philadelphia: Francis and Robert Bailey, 1800), 16.

110. Edgar P. Richardson, Brooke Hindle, and Lillian B. Miller, *Charles Willson Peale and His World* (New York: Abrams, 1983), 87, 123.

111. C. Sellers, *Mr. Peale's Museum*, 202. Elizabeth Drinker recorded a visit by her daughter and granddaughter in 1795: "Jacob [Downing], Sally and their Daughters were here this afternoon, they wish'd me to go with them to the Statehouse Yard, and to Peels Museum which is keep't in the State House, I declin'd the motion, but Molly excepted it, and went with them." Drinker, *Diary*, 1:686, entry for May 30, 1795.

112. Thomas Jefferson, *Notes on the State of Virginia*, 1787 (University of Virginia Library, Electronic Text Center), 179, accessed March 12, 2019, http://etext.lib. virginia.edu/toc/modeng/public/JefVirg.html. Jefferson first refuted claims about America in *Notes on the State of Virginia*. In answer to a published comment by Abbé Raynal that America had not yet produced one good poet, mathematician, or genius in any art or science, Jefferson listed several:

> Washington, whose memory will be adored while liberty shall have vota-
> ries, whose name will triumph over time, and will in future ages assume its
> just station among the most celebrated worthies of the world, when that
> wretched philosophy shall be forgotten which would have arranged him
> among the degeneracies of nature. In physics we have produced a Franklin,
> than whom no one of the present age has made more important discover-
> ies, nor has enriched philosophy with more, or more ingenious solutions of
> the phaenomena of nature. We have supposed Mr. Rittenhouse second to no
> astronomer living. that in genius he must be the first, because he is self-taught.
> As an artist he has exhibited as great a proof of mechanical genius as the world
> has ever produced. He has not indeed made a world; but he has by imitation
> approached nearer its Maker than any man who has lived from the creation to
> this day. (*Notes on the State of Virginia*, 190–91)

113. Jefferson, *Notes on the State of Virginia*, 190.

114. *State Gazette of South Carolina* (Charleston), July 29, 1793. Cited in Lee Alan Dugatkin, *Mr. Jefferson and the Giant Moose: Natural History in Early America* (Chicago: University of Chicago Press, 2009), 16.

115. Jefferson, *Notes on the State of Virginia*, Query 6, 177, quoted in Dustin Gish and Daniel Klinghard, *Thomas Jefferson and the Science of Republican Government: A Political Biography of Notes on the State of Virginia* (Cambridge: Cambridge University Press, 2017), 145.

116. "The Utility of the Union in Respect to Commercial Relations and a Navy," *Independent Journal* (New York), November 24, 1787. Gish and Klinghard, *Thomas Jefferson*, 139–40.

117. "The Mammoth, or Big Buffalo," *Massachusetts Gazette* (Boston), May 4, 1787; *New Haven Gazette and Connecticut Magazine*, May 3, 1787; *Pennsylvania Packet, and Daily Advertiser* (Philadelphia), June 13, 1787.

118. Peale advertised the opening of the mammoth exhibit in *Poulson's American Daily Advertiser* (Philadelphia), December 30, 1801. Drinker, *Diary*, 3:1475–76, entry for December 30, 1801.

119. C. Sellers, *Mr. Peale's Museum*, 142.

120. Deborah Norris Logan to Albanus Logan, January 10, 1802, Robt R. Logan Collection, Historical Society of Pennsylvania, quoted in David R. Brigham, *Public Culture in the Early Republic: Peal's Museum and Its Audience* (Washington, DC: Smithsonian Institution Press, 1995), 65.

121. Daniel L. Dreisbach, "Mr. Jefferson, a Mammoth Cheese, and the 'Wall of Separation between Church and State': A Bicentennial Commemoration," *Journal of Church and State* 43, no. 4 (Autumn 2001): 725.

122. "The Mammoth Cheese. An Epico-Lyrico Ballad," *Connecticut Courant* (Hartford), September 14, 1801.

123. "The Great Cheese," *Hampshire Gazette* (Northampton, MA), September 30, 1801.

124. Don Federo, "The Demo' Mammo' Cheese. An Epic Poem," *Albany (NY) Centinel*, August 14, 1801.

125. "What Is Cheese, without Its Equivalent? a Mammoth Pye,-Proposed," *Ploughman; or, Republican Federalist* (Bennington, VT), September 21, 1801.

126. "A Mammoth Beet," *Norwich (CT) Packet*, November 10, 1801.

127. *Gazette of the United States* (Philadelphia), August 8, 1801; *New-York Gazette and General Advertiser*, December 7, 1801; "The Mammoth Cheese Afloat!," *Independent Gazetteer* (Worcester, MA), December 15, 1801. *Poulson's American Daily Advertiser* (Philadelphia) reported the cheese arrived in Washington in a wagon drawn by six horses (January 4, 1802).

128. Martha Washington's granddaughter Eleanor Custis to Mary Pinckney, January 3, 1802, quoted in C. A. Browne, "Elder John Leland and the Mammoth Cheshire Cheese," *Agricultural History* 18, no. 4 (October 1944): 151.

129. *New York Evening Post* article report in the *Republican or, Anti-Democrat* (Baltimore), January 5, 1802.

130. *New York Evening Post* article report in the *Republican or, Anti-Democrat*. Jeffrey Pasley discussed the political overtones of the mammoth cheese in "The Cheese and the Words: Popular Political Culture and Participatory Democracy in the Early American Republic," in *Beyond the Founders: New Approaches to the Political History of the Early American Republic*, ed. Jeffrey L. Pasley, Andrew W. Robertson, and David Waldstreicher (Chapel Hill: University of North Carolina Press, 2004), 31–54.

131. Semonin, *American Monster*, 361. Semonin refers to Laura Rigal's analysis of the painting, "Peale's Mammoth," in *American Iconology: New Approaches to Nineteenth-Century Art and Literature*, ed. David C. Miller (New Haven, CT: Yale University Press, 1993), 18–38.

132. "From Benjamin Franklin to Sir John Pringle, [before May 10, 1772]," Founders Online, National Archives, accessed September 29, 2019, https://found ers.archives.gov/documents/Franklin/01-19-02-0096. (Original source: *The Papers of Benjamin Franklin*, vol. 19, *January 1 through December 31, 1772*, ed. William B. Willcox [New Haven, CT: Yale University Press, 1975], 139.)

133. Anishanslin, *Portrait of a Woman in Silk*, 308. Bache wrote, "It will shew what can be sent from America to the looms of France." "To Benjamin Franklin from Sarah Bache, 14 September 1779," Founders Online, National Archives, accessed September 29, 2019, https://founders.archives.gov/documents/Franklin/01-30-02-0271. (Original source: *The Papers of Benjamin Franklin*, vol. 30, *July 1 through October 31, 1779*, ed. Barbara B. Oberg [New Haven, CT: Yale University Press, 1993], 332–36.)

2. Flights of Imagination

1. Benjamin Rush to Elizabeth Graeme Ferguson, January 18, 1793, in *Letters of Benjamin Rush*, vol. 2, *1793–1813*, ed. L. H. Butterfield (Princeton, NJ: Princeton University Press, 1951), 627.

2. Benjamin Franklin Bache diary, August 1, 1782—September 14, 1785, entry for Tuesday, March 2, 1784, Bache Collection of Franklin Papers, American Philosophical Society. English translation of the original diary Bache kept in French.

3. Bache diary, entry for Wednesday, June 23, 1784. Described in L. T. C. Rolt, *The Aeronauts: A History of Ballooning, 1783–1903* (New York: Walker, 1966), 58. In July, Bache traveled with Franklin to Saint-Cloud, a suburb of Paris, to see an ascension. Bache wrote: "It went up with four persons, among others the Duke de Chartres. They were soon lost to view in the clouds which were very thick, and where they found it very tempestuous, but they soon arose above them and found the Sun which so expanded the air in their Balloon, that they could not open the valve, so they were obliged to break it, and they fell about two miles from their point of departure." Bache diary, entry for Thursday, July 15, 1784.

4. Bache diary, entry for Sunday, September 19, 1784.

5. The Adams family resided in Auteuil outside Paris during 1784–85. Abigail A. Smith and C. A. S. De Windt, *Journal and Correspondence of Miss Adams, Daughter of John Adams, Second President of the United States: Written in France and England in 1785*, vol. 1 (New York: Wiley and Putnam, 1841), 19.

6. "They left Dover with their Balloon already full of holes; after having been for an hour over the sea, their Balloon losing its gas all the time, fell considerably. This accident forced them to throw out all their Ballast, this happening several times they were obliged to throw out, first all the ornaments of their car, namely the printed cloth and the garlands which ornamented it, secondly, all their clothes except their shirts and Mr. Jeffries cork packet, with which he had provided himself, to be able to float, in case they came to this extremity (and they thought really it had come; but Mr. Blanchard opposed it) and in two hours and three quarters, they came down on

the top of a tree, where they remained twenty eight minutes in their shirts; thence to Callais [*sic*], where they were carried as in triumph." Bache diary, entry for January 1785.

7. Smith and De Windt, *Journal and Correspondence of Miss Adams*, 48. Abigail Adams also recorded the details of Blanchard and Jeffries's journey across the Channel:

> Last Friday, the 7th of January, Mr. Blanchard and Dr. Jeffries ascended at Dover in a balloon, and in two hours descended a league from Calais, to the great joy and admiration of every one who saw them. The people of Calais received the aeriel travellers with every mark of attention, respect, and admiration; they presented Mr. Blanchard with a gold box, the figure of his balloon on the cover, and presented him with letters, giving him the title of citizen of Calais. They offered the same to Dr. J. but he, being a stranger, declined them; probably thinking his situation in England would be rendered more disagreeable, and create jealousies by such a distinction. They likewise requested of Mr. Blanchard his balloon to put into the Cathedral Church at Calais, as the ship of Columbus was put into a Church in Spain. These gentlemen have arrived at Paris. This voyage has been long projected—their success has been quite equal to their expectations; there being but little wind, they did not make so quick a voyage as some others have done. Mr. B. is a Frenchman, Dr. J. an American. (40–41)

8. Bache diary, entry for Sunday, July 11, 1784.

9. Jonathan Williams Jr. to Benjamin Franklin, London, May 14, 1785, Papers of Benjamin Franklin, accessed April 2, 2020. https://franklinpapers.org/framed Volumes.jsp.

10. Franklin to Henry Laurens, Passy, December 6, 1783, Papers of Benjamin Franklin.

11. Franklin to Jan Ingenhousz, Passy, January 16, 1784, Papers of Benjamin Franklin.

12. Abigail Adams to Lucy Cranch, June 23, 1785, Founders Online, National Archives, http://founders.archives.gov/documents/Adams/04-06-02-0064.

13. Daughter Abigail married June 11. Adams lamented the loss of her companionship in England. Abigail Adams to Mary Smith Cranch, June 13, 1786, Founders Online, National Archives, last modified December 28, 2016, http://founders.archives.gov/documents/Adams/04-07-02-0081.

14. "Extracts from the Paris Journals of August 28 and 29, giving an account of the AIR BALLOONS thrown up at Paris," *Salem (MA) Gazette*, November 27, 1783.

15. "Copy of a Letter Addressed to a Former Secretary of the American Philosophical Society," *New-Haven (CT) Gazette*, July 8, 1784.

16. *Boston Magazine*, February 1784; "Explanation of the Air Balloon," *Boston Magazine*, May 1784.

17. *Weatherwise's Town and Country Almanac for 1785* (Boston: Weeden and Barrett, 1784).

18. Michael E. Connaughton, "'Ballomania': The American Philosophical Society and Eighteenth-Century Science," *Journal of American Culture* 7 (Spring/Summer 1984): 72.

19. Francis Hopkinson, Philada, May 24, 1784, Papers of Benjamin Franklin.

20. Thomas Jefferson from Francis Hopkinson, May 12, 1784, Founders Online, National Archives, http://founders.archives.gov/documents/Jefferson/01-07-02-0202.

21. "Pd. for 2 tickets to see balon 15/.33," Founders Online, National Archives, http://founders.archives.gov/documents/Jefferson/02-01-02-0018. Jefferson did not depart for France until July 1784. He wrote to his friend James Monroe, "I have had the pleasure of seeing three balons here. The largest was 8.f. diameter, and ascended about 300 feet." Thomas Jefferson to James Monroe, May 21, 1784, Founders Online, National Archives, accessed March 26, 2017, http://founders.archives.gov/documents/Jefferson/01-07-02-0226.

22. George Washington to Duportail, April 4, 1784, Founders Online, National Archives, http://founders.archives.gov/documents/Washington/04-01-02-0189.

23. George Washington to Edward Newenham, November 25, 1785, Founders Online, National Archives, http://founders.archives.gov/documents/Washington/04-03-02-0346.

24. Shortly after this, on December 8, the *Virginia Journal and Alexandria Advertiser* printed a news item: "Richmond, December 3. On Saturday last, between four and five o'clock in the afternoon, Mr. Busselot, a French gentleman, raised a Balloon from the capitol square on Shockoe-Hill, in this city, which ascended to a great height. The wind setting North-east, it took that course, and descended before night ten miles distant from the city, on the plantation of Captain John Austin, in Hanover." George Washington to Edward Newenham, November 25, 1785, footnote 2, Founders Online, National Archives, http://founders.archives.gov/documents/Washington/04-03-02-0346.

25. Rev. James Madison to Thomas Jefferson, April 28, 1784, Founders Online, National Archives, http://founders.archives.gov/documents/Jefferson/01-07-02-0140.

26. Rev. James Madison to Thomas Jefferson, March 27, 1786, Founders Online, National Archives, http://founders.archives.gov/documents/Jefferson/01-09-02-0314.

27. Tom D. Crouch, *The Eagle Aloft: Two Centuries of the Balloon in America* (Washington, DC, Smithsonian Institution Press, 1983), 66. The June 24 flight of thirteen-year-old Edward Warren was reported in the *Maryland Journal and Baltimore Advertiser*, June 26, 1784.

28. *Pennsylvania Packet, and General Advertiser* (Philadelphia), June 12, 1784.

29. Some spectators, who had not seen Carnes tumble from the basket, assumed the worst: "on the supposition of a person having gone up with the *balloon*; and their apprehensions were increased by the falling of the furnace, which, to those not near, presented to their imaginations the dreadful spectacle of a man falling from an immense height." *Pennsylvania Gazette* (Philadelphia), July 21, 1784.

30. *Massachusetts Centinel and the Republican Journal* (Boston), June 30, 1784.

31. *Maryland Journal and Baltimore Advertiser*, June 22, 1784, quoted in Crouch, *Eagle Aloft*, 69.

32. Francis Hopkinson to Thomas Jefferson, March 31, 1784, Founders Online, National Archives, https://founders.archives.gov/documents/Jefferson/01-07-02-0056.

33. *American Herald* (Boston), July 5, 1784.

34. *South-Carolina Gazette and Public Advertiser* (Charleston), October 13, 1784.

35. *Independent Gazetteer* (Philadelphia), August 7, 1784; *Pennsylvania Packet, and Daily Advertiser* (Philadelphia), November 2, 1784.

36. United States Patent Office, *A List of Patents Granted by the United States from April 10, 1790, to December 31, 1836: With an Appendix Containing Reports on the Condition of the Patent-Office in 1823, 1830, and 1831* (Washington, DC, 1872), 20.

37. *Newport (RI) Mercury*, July 24, 1784; *New York Packet and American Advertiser*, July 15, 1784; *Independent Chronicle and Universal Advertiser* (Boston), September 30, 1784; *American Herald* (Boston), July 19, 1784; *Freeman's Journal: or, The North-American Intelligencer* (Philadelphia), September 15, 1784; *Massachusetts Centinel and the Republican Journal* (Boston), June 16, 1784; *New-Hampshire Gazette and General Advertiser* (Portsmouth), May 29, 1784; *Virginia Journal and Alexandria Advertiser*, June 17, 1784; *Independent Gazetteer* (Philadelphia), July 14, 1784.

38. *American Herald* (Boston), July 19, 1784; *South-Carolina Gazette and General Advertiser* (Charleston), January 10, 1785.

39. *Independent Ledger and the American Advertiser* (Boston), July 5, 1784.

40. "Fashion," *Political Intelligencer and New-Jersey Advertiser* (New Brunswick), October 12, 1784.

41. "A Balloon Song," *Political Intelligencer and New-Jersey Advertiser* (New Brunswick), December 28, 1784.

42. "Balloon influenza" under "Foreign Intelligence," *New-Jersey Gazette* (Trenton), November 22, 1784; *Pennsylvania Gazette* (Philadelphia), July 28, 1784. Repeated in *Massachusetts Centinel and the Republican Journal* (Boston), July 14, 1784.

43. "Balloon Wish," broadside (Boston: printed by Powars and Willis, 1784).

44. "Kingston, (Jamaica) April 6," *Connecticut Journal* (Hartford), July 6, 1785.

45. Philip Freneau, "The Progress of Balloons," *Freeman's Journal: or, The North-American Intelligencer* (Philadelphia), December 22, 1784.

46. *Falmouth (ME) Gazette and Weekly Advertiser*, April 2, 1785.

47. *An Account of Count D'Artois and his Friend's Passage to the Moon in a Flying Machine, called, An Air Balloon* (Litchfield, CT: Collier and Capp, 1785).

48. James West, *The New-York Balloon; or, A Mogul Tale* (1797), advertised in the *Diary or Loudon's Register* (New York), March 11, 1797.

49. *Freeman's Journal; or, The North-American Intelligencer* (Philadelphia), July 29, 1789.

50. *New York Journal*, August 1789.

51. *Pennsylvania Packet, and General Advertiser* (Philadelphia), June 29, 1784; *Independent Gazetteer* (Philadelphia), June 26 and July 29, 1784.

52. "From the Pennsylvania Journal," *American Herald* (Boston), April 5, 1784.

53. Diary of John Quincy Adams, Founders Online, National Archives, http://founders.archives.gov/documents/Adams/03-01-02-0007-0003.

54. Jefferson wrote to his daughter: "We were entertained here lately with the ascent of Mr. Blanchard in a *baloon*. The security of the thing appeared so great that every body is wishing for a baloon to travel in. I wish for one sincerely, as instead of 10. days, I should be within 5 hours of home. Maria will probably give you the baloon details, as she writes to-day." Thomas Jefferson to Martha Jefferson Randolph, January 14, 1793, Founders Online, National Archives, http://founders.archives.gov/documents/Jefferson/01-25-02-0056.

55. "The President took off his hat, and bowed to him as he ascended." *Federal Gazette and Philadelphia Daily Advertiser*, January 9, 1793.

56. Benjamin Rush, *The Autobiography of Benjamin Rush: His "Travels through Life" together with His Commonplace Book for 1789–1813*, ed. George W. Corner (Princeton, NJ: Princeton University Press, 1948), 304.

57. *The Diaries of Samuel Mickle: Woodbury, Gloucester County, New Jersey, 1792–1829*, transcribed by Ruthe Baker (Woodbury, NJ: Gloucester County Historical Society, 1991), 7, entry for January 8, 1793.

58. Rush, *Autobiography*, 304.

59. "Aerostation. Philadelphia, Dec. 24," *Salem (MA) Gazette*, January 8, 1793.

60. "The Camden Balloon," *Boston Daily Advertiser*, September 9, 1819.

61. "The Diary of Samuel Breck, 1814–1822," ed. Nicholas B. Wainwright, *Pennsylvania Magazine of History and Biography* 102, no. 4 (October 1978): 497, diary entry for September 8, 1819.

62. "Another Disappointment!," *Baltimore Patriot and Mercantile Advertiser*, September 10, 1819.

63. Franklin to Jan Ingenhousz, Passy, January 16, 1784, Papers of Benjamin Franklin.

64. *Baltimore Gazette and Daily Advertiser*, July 28, 1828.

65. "A Peep into Futurity," *Farmers' Cabinet* (Amherst, NH), March 20, 1851.

66. "J. V. Cameron & Co.'s Dramatic Company!," playbill (Fitchburg, MA, 1854), American Antiquarian Society.

67. Matthew Goodman, *The Sun and the Moon: The Remarkable True Account of Hoaxers, Showmen, Dueling Journalists, and Lunar Man-Bats in Nineteenth-Century New York* (New York: Basic Books, 2008), 238.

68. Thomas Ollive Mabbott (E. A. Poe), "The Balloon Hoax," in *The Collected Works of Edgar Allan Poe*, vol. 3 (Cambridge, MA: Harvard University Press, 1978), 1065, accessed June 21, 2017, https://www.eapoe.org/works/mabbott/tom3t015.htm.

69. Monck Mason, *Account of the late Aeronautical Expedition from London to Weilburg* (New York: T. Foster, 1837); *Remarks on the Ellipsoidal Balloon, Propelled by the Archimedean Screw, Described as the New Aerial Machine* (London: Howlett and Son, 1843); Ronald Sterne Wilkinson, "Poe's 'Balloon-Hoax' Once More," *American Literature* 32, no. 3 (November 1960): 313–17.

70. Edgar Allan Poe, "The Unparalleled Adventure of One Hans Pfaall," *Southern Literary Messenger* 1, no. 10 (June 1835): 565–80.

71. *New York Sun*, April 15, 1844; Goodman, *Sun and the Moon*, 244–45.

72. Peter Parley, "About Balloons," *Parley's Magazine for Children and Youth* (Boston: Lilly, Wait), August 31, 1833, 85, 88. Cover illustration with caption "About Balloons" shows a man sinking in waves with his balloon, possibly a depiction of the Englishman Major Money who descended in the Channel and was rescued by a passing ship.

73. "Mr. Durant," by James ("a Young Correspondent"), *Parley's Magazine*, September 14, 1833, 18. The following year *Parley's Magazine* addressed the question, "Why Cannot Men Fly?" The author explained why men were different from birds but offered speculation that technology might allow a person to fly. The article was illustrated with an image of a man wearing artificial wings and a small balloon. Peter Parley, *Parley's Magazine*, December 20, 1834, 5–6.

74. Mary A. Swift, *First Lessons on Natural Philosophy for Children, Part Second* (Hartford, CT: Belknap and Hamersley, 1836), 103–4. The back page has reviews of

the book *New-Haven Palladium*: "Its contents are admirably adapted to their capacities, the science being illustrated by the things most familiar to their sight and understanding."

75. Theodore Thinker, *The Balloon and Other Stories* (New York: Clark, Austin, 1851).

76. Samuel Colman, *The Child's Gem* (New York: S. Colman, 1839).

77. *Aunt Mary's Stories for Children* (New York: Leavitt and Allen, 1852).

78. Samuel G. Goodrich, *Robert Merry's Museum*, vol. 6 (New York: G. W. and S. O. Post, 1847).

79. "Merry's Balloon Travels" was published serially in *Robert Merry's Museum* in 1852, then as a book in 1856: Peter Parley, ed., *Balloon Travels of Robert Merry and His Young Friends over Various Countries in Europe* (New York: J. C. Derby, 1856).

80. "The Balloon," in *The Evergreen: or, Stories for Childhood and Youth*, ed. Walter West (Boston: Munroe and Francis, 1847), 15–20.

81. Advertisement for Munroe and Francis publications at the back of the *Evergreen*. E. Landell, *The Boy's Own Toy-Maker: A Practical Illustrated Guide to the Useful Employment of Leisure Hours* (Boston: Shepard, Clark and Brown, 1859), included plans for making kites and small fire balloons.

82. Francis Hopkinson to Benjamin Franklin, Philadelphia, May 24, 1784, Papers of Benjamin Franklin, accessed June 25, 2017, http://franklinpapers.org/franklin//framedNames.jsp.

83. *Dunlap's American Daily Advertiser* (Philadelphia), February 1, 1793.

84. Milo M. Naeve, *John Lewis Krimmel: An Artist in Federal America* (Newark: University of Delaware Press, 1988), 116–17.

85. Naeve, *John Lewis Krimmel*, 117; *Poulson's American Daily Advertiser* (Philadelphia), March 13, 1821.

86. *Pennsylvania Gazette* (Philadelphia), March 16, 1821. A cattle show in Pawtuxet, Rhode Island, 1831, advertised not one but two balloon launches as part of the activities. *Rhode Island American, Statesman and Providence Gazette*, September 27, 1831.

87. Quoted in Charlene Mires, "Lafayette's Tour," Encyclopedia of Greater Philadelphia, accessed October 10, 2017, http://philadelphiaencyclopedia.org/archive/lafayettes-tour/.

88. *Baltimore Patriot and Mercantile Advertiser*, September 14, 1824.

89. Lafayette's Arrival at Independence Hall, Germantown Print Works, 1824–25, Germantown, Philadelphia, Winterthur Museum, 1967, 144, http://philadelphiaencyclopedia.org/archive/lafayettes-tour/#21033. The balloon was launched by Messrs. Brown and Regnault, who also staged a fireworks exhibition in honor of Lafayette on the evening of his arrival in the city. "Exhibition of Fire Works, in honor of La Fayette," *Aurora General Advertiser* (Philadelphia), September 27, 1824.

90. *New-Hampshire Sentinel* (Keene), October 8, 1830. Durant claimed to be the "first American whoever ascended in a balloon alone." He was not. "The aeronaut of the people! Distributed by Mr. Charles F. Durant, on his ninth grand ascension. Observatory Garden, (Federal Hill), October 14th, 1833," broadside (1833), MS-337, Richard Gimbel Collection, United States Air Force Academy archives.

91. Crouch cites numerous newspaper reports for September 9–14, 1830. Crouch, *Eagle Aloft*, 147.

92. William Nester, *The Age of Jackson and the Art of American Power, 1815–1848* (Washington, DC: Potomac Books, 2013), 153–55.

93. *Eastern Argus* (Portland, ME), June 17, 1833.

94. *Richmond (VA) Enquirer*, July 16, 1833.

95. Samuel George Morton, *Crania Americana; or, A Comparative View of the Skulls of Various Aboriginal Nations of North and South to Which Is Prefixed an Essay on the Varieties of the Human Species* (Philadelphia: J. Dobson, 1839), 6. Morton sent out a prospectus asking for Native American skulls. Many of his respondents supplied him with skulls robbed from graves. Ann Fabian, *The Skull Collectors: Race, Science, and America's Unburied Dead* (Chicago: University of Chicago Press, 2010), 36–38; Bruce R. Dain, *A Hideous Monster of the Mind: American Race Theory in the Early Republic* (Cambridge, MA: Harvard University Press, 2002), 199.

96. "Phrenological Developments and Character of the Celebrated Indian Chief and Warrior, Black Hawk; With Cuts," *American Phrenological Journal and Miscellany*, November 1, 1838, 53–54.

97. Cass, who prior to becoming secretary of war was superintendent of Indian affairs in Michigan, "argued that Indian languages were ugly, grammatically primitive, and incapable of expressing abstract ideas, all of which Cass saw as proof that the potential range of Indians' thoughts was so limited that civilizing efforts were destined to fail." Cameron Strang, "Scientific Instructions and Native American Linguistics in the Imperial United States: The Department of War's 1826 Vocabulary," *Journal of the Early Republic* 37 (Fall 2017): 411–12. Cass's call for extermination is quoted in Willard Carl Klunder, *Lewis Cass and the Politics of Moderation* (Kent, OH: Kent State University Press, 1996), 10.

98. President Jackson's Message to Congress, "On Indian Removal," December 6, 1830, Record Group 46, Records of the United States Senate, 1789–1990, National Archives and Records Administration (NARA), Washington, DC. Six months after the captives' eastern tour, Jackson articulated the racism behind his policy of Indian removal:

> My original convictions upon this subject have been confirmed by the course of events for several years, and experience is every day adding to their strength. That those tribes can not exist surrounded by our settlements and in continual contact with our citizens is certain. They have neither the intelligence, the industry, the moral habits, nor the desire of improvement which are essential to any favorable change in their condition. Established in the midst of another and a superior race, and without appreciating the causes of their inferiority or seeking to control them, they must necessarily yield to the force of circumstances and ere long disappear.

President Andrew Jackson, Fifth Annual Message to Congress, December 3, 1833, in American Presidency Project, accessed June 13, 2017, http://www.presidency. ucsb.edu/ws/?pid=29475.

99. Kerry A. Trask, *Black Hawk: The Battle for the Heart of America* (New York: Henry Holt, 2006), 301.

100. "Address to Black Hawk," *Salem (MA) Gazette*, June 25, 1833.

101. *Eastern Argus* (Portland, ME), June 17, 1833.

102. *Eastern Argus* (Portland, ME), June 19, 1833.

103. *Farmer's Cabinet* (Amherst, NH), June 21, 1833.

104. *Salem (MA) Gazette*, June 18, 1833.

105. David Claypoole Johnston, "The Grand National Caravan Moving East" (New York: Endicott and Swett, 1833).

106. Erika Piola and Jennifer Ambrose, "The First Fifty Years of Commercial Lithography in Philadelphia," in *Philadelphia on Stone: Commercial Lithography in Philadelphia, 1828–1878*, ed. Erika Piola (University Park: Penn State University Press, 2012), 1–47.

107. Edward Williams Clay, "The Times," lithograph (New York: H. R. Robinson, 1837), Library of Congress; Jessica M. Lepler, *The Many Panics of 1837: People, Politics, and the Creation of a Transatlantic Financial Crisis* (Cambridge: Cambridge University Press; 2013). Lepler points out that the image was made post-panic (sometime after May 1837) and is really "an argument about panic's political causes," 251. Nancy Reynolds Davison, "E. W. Clay: American Political Caricaturist of the Jacksonian Era" (PhD diss., University of Michigan, 1980). The air balloon marked "Safety Fund" refers to Van Buren's insurance program for New York banks. Davison speculates the date of printing is July 1837, 160.

108. H. Bucholzer, "Balloon ascension to the Presidential Chair" (New York: Lith. & pub. by James Baillie, 1844); H. Bucholzer, "Bursting the Balloon" (New York: Lith. & pub. by James Baillie, 1844), deposited in Library of Congress June or early July 1844. Thomas C. Blaisdell and Peter Selz, *The American Presidency in Political Cartoons, 1776–1976* (Berkeley, CA: University Art Museum, 1976), 74; "Grand Whig Procession at Louisville," *Farmer's Cabinet* (Amherst, NH), August 1, 1844.

109. *Hudson River Chronicle* (Ossining, NY), August 20, 1844.

110. *Barre (MA) Gazette*, August 23, 1844.

111. Chronology in Jean Lipman, *Rufus Porter Rediscovered: Artist, Inventor, Journalist, 1792–1884* (New York: C. N. Potter, 1980), 187.

112. Circulation was between three thousand and four thousand during Porter's time as editor. Crouch, *Eagle Aloft*, 295. Masthead reproduced in Lipman, *Rufus Porter*, 52. The full title was *New York Mechanic: The Advocate of Industry and Enterprise, and Journal of Mechanical, and Other Scientific Improvements.*

113. *Scientific American* first issue, Rufus Porter, "To the American Public," August 28, 1845, 2. The first issue informed readers what the magazine contained and who its primary audience was:

> Each number will be furnished with from two to five original Engravings, many of them elegant, and illustrative of New Inventions, Scientific Principles, and Curious Works; and will contain, in addition to the most interesting news of passing events, general notices of progress of Mechanical and other Scientific Improvements; American and Foreign. Improvements and Inventions; Catalogues of American Patents; Scientific Essays, illustrative of the principles of the sciences of Mechanics, Chemistry, and Architecture: useful information and instruction in various Arts and Trades; Curious Philosophical Experiments; Miscellaneous Intelligence, Music and Poetry. This paper is especially entitled to the patronage of Mechanics and Manufactures, being the only paper in America, devoted to the interest of those classes; but is particularly useful to farmers, as it will not only appraise them of improvements in agriculture implements, but instruct them in various mechanical trades, and guard them against impositions.

Porter sold the magazine in 1847. He then edited and published the *Scientific Mechanic* from 1847 to 1848. Lipman, *Rufus Porter*, 58.

114. *Scientific American*, September 18, 1845, 1. The December 25, 1847, issue carried an illustration of the airship on the front page.

115. Rufus Porter, *Aerial Navigation: The Practicality of Traveling Pleasantly and Safely from New-York to California in Three Days* (New York: H. Smith, 1849), 16. In 1851 Porter petitioned the US Senate for funds (he was denied). Crouch, *Eagle Aloft*, 311. Porter subsequently organized a joint stock company, the Aerial Navigation Company, and began to build the airship. The following year Porter edited and published a bimonthly newsletter, the *Aerial Reporter* (Washington, DC), designed to promote the airship. In 1853 Porter exhibited a twenty-foot working model of the airship at Carusi's Hall in Washington, DC. Lipman, *Rufus Porter*, 60, 188. The full-size ship was never completed.

116. "The Travelling Balloon," *Scientific American*, March 3, 1849, 189, 188, 185.

117. "The New Flying Ship," *Gleason's Pictorial Drawing-Room Companion*, August 16, 1851, 256. Porter had gone into partnership with a Mr. Robjohn. He, and not Porter, is mentioned in the article. Crouch, *Eagle Aloft*, 310–11.

118. "A Fruitless Application," *Aerial Reporter*, July 31, 1852, 2, cited in Crouch, *Eagle Aloft*, 311.

119. "The Great Pictorial Romance of the Age, or Steam Ship Commodores & United States Mail Contractors." Colonel Richard Gimbel Aeronautical History Collection, United States Air Force Academy, McDermott Library, XL-28 2270.

120. "The Way They Go to California," lithograph (Spruce, NY: N. Currier, c. 1849), Library of Congress Prints and Photographs Division, http://www.loc.gov/pictures/resource/pga.05072.

121. Crouch, *Eagle Aloft*, 357.

122. Crouch, 354–55, 357.

123. Crouch, 356. The Confederacy made only two attempts to use balloons for aerial observation. Both balloons were destroyed or captured by the Union army. Crouch, *Eagle Aloft*, 392–93.

124. Civil War envelope collection, American Antiquarian Society, Worcester, MA.

125. *Harper's Weekly* (New York), July 27, 1861, 476. There were over three hundred reporters from Northern newspapers and magazines in the field from 1861 to 1865. J. Cutler Andrews, *The North Reports the Civil War* (Pittsburgh: University of Pittsburgh Press, 1955), 60. Currier and Ives, "Battle of Fair Oaks, Va. May 31 1862," lithograph, Smithsonian National Air and Space Museum, object number A20140397000.

126. *New York Times*, September 9, 1861; "War Balloon at General McDowell's Headquarters," *Harper's Weekly*, October 26, 1861.

127. *New York Times*, June 5, 1862.

128. *New York Times*, May 24, 1862.

129. Andrews, *North Reports the Civil War*, 198.

130. *Frank Leslie's Illustrated Newspaper* (New York), February 22, 1862; Ford Risley, *Civil War Journalism* (Santa Barbara, CA: Praeger, 2012), 79–80; Kristen M. Smith and Jennifer L. Gross, *The Lines Are Drawn: Political Cartoons of the Civil War* (Athens, GA: Hill Street, 1999), 60.

131. Eugene B. Block, *Above the Civil War: The Story of Thaddeus Lowe, Balloonist, Inventor, Railway Builder* (Berkeley, CA: Howell-North, 1966), 56.

132. *New York Times*, June 15, 1862.

133. *New York Times*, June 15, 1862; Charles M. Evans, *The War of the Aeronauts: A History of Ballooning during the Civil War* (Mechanicsburg, PA: Stackpole Books, 2002), 166.

134. Crouch, *Eagle Aloft*, 336.

135. Stanley and Conant's *Polemorama, or Gigantic Illustrations of the War*, broadside 22855, American Antiquarian Society.

136. "Lieut. Genl. Ulysses S. Grant, U.S.A." (New York: J. C. Buttre, 48 Franklin St., New York, 1864), American Antiquarian Society, 446312.

137. "National Ode" (Boston: J. E. Farwell, 1864), Broadside 79, American Antiquarian Society.

138. *Rhode Island American, Statesman and Providence Gazette*, September 18, 1829.

3. Engines of Change

1. *Federal Gazette and Philadelphia Daily Advertiser*, January 22, 1789.

2. "Mr. R has the most perfect conviction that the motion of the machine will never cease so long as the materials of which it is composed will last. He has exhibited it to many of his neighbors, all of whom express their astonishment at the perpetuity of its motion." *City Gazette and Daily Advertiser* (Charleston, SC), April 28, 1812.

3. Simon Schaffer, "The Show That Never Ends: Perpetual Motion in the Early Eighteenth Century," *British Journal for the History of Science* 28, pt. 2, no. 97 (June 1995): 157–89; Henry Dircks, *Perpetuum Mobile; or, A History of the Search for Self-Motive Power, from the 13th to the 19th Century* (London: E. & F. N. Spon., 1861; repr., Amsterdam: B. M. Israël, 1968).

4. "Discovery of the Perpetual Motion!," *Columbian Centinel* (Boston), February 25, 1795.

5. *New York Evening Post*, April 14, 1802; Rita Susswein Gottesman, *Arts and Crafts in New York, 1800–1804: Advertisements and News Items from New York City Newspapers* (New York: New-York Historical Society, 1965), 405.

6. *Spectator* (New York), February 18, 1800; Gottesman, *Arts and Crafts in New York*, 404.

7. "Perpetual Motion," *Independent Gazetteer* (Philadelphia), November 25, 1795.

8. David F. Launy to Thomas Jefferson, January 1, 1805, Founders Online, National Archives, http://founders.archives.gov/documents/Jefferson/99-01-02-0922.

9. Matthew Wilson to Thomas Jefferson, March 18, 1805, Founders Online, National Archives, http://founders.archives.gov/documents/Jefferson/99-01-02-1406.

10. "Sir ibeing in low circumstance idesire you to send mee an order which will answer my expences to Washington as iam determined not to discover it untill isee you personally." Ambrose Bayley to Thomas Jefferson, May 1, 1806, Founders Online, National Archives, http://founders.archives.gov/documents/Jefferson/99-01-02-3672.

11. Joseph O'Neil to Thomas Jefferson, March 2, 1809, Founders Online, National Archives, http://founders.archives.gov/documents/Jefferson/99-01-02-9941.

12. Dayton Leonard to James Madison, December 21, 1810 (abstract), Founders Online, National Archives, http://founders.archives.gov/documents/Madison/03-03-02-0095.

13. "Perpetual Motion," *City Gazette and Daily Advertiser* (Charleston, SC), November 14, 1801; *American Citizen and General Advertiser* (New York), December 23, 1801; Gottesman, *Arts and Crafts in New York*, 404.

14. *Proceedings and Debates of the House of Representatives of the United States* (known as the *Annals of Congress*), 1st Cong., 2nd Sess. (1802), 376. http://lcweb2.loc.gov/ammem/amlaw/lwac.html.

15. Lewis Du Pré to Thomas Jefferson, February 7, 1802, Founders Online, National Archives, http://founders.archives.gov/documents/Jefferson/01-36-02-0338.

16. *Proceedings and Debates of the House of Representatives of the United States* (known as the *Annals of Congress*), 1st Cong., 2d Sess. (1802), 376–77, 470. http://lcweb2.loc.gov/ammem/amlaw/lwac.html.

17. "For the Portfolio," *Port Folio* (Philadelphia), January 23, 1802, 24.

18. "Valuable Secrets," *The Columbian Almanac, or The North American Almanac*, 1804 (Wilmington, DE, 1803).

19. "New and Pleasing Entertainment," *Repertory* (Boston), December 26, 1806.

20. *City Gazette and Daily Advertiser* (Charleston, SC), April 28, 1812.

21. *Philadelphia Merchant: The Diary of Thomas P. Cope, 1800–1851*, ed. Eliza Cope Harrison (South Bend, IN: Gateway Editions, 1978), 277.

22. *Aurora General Advertiser* (Philadelphia), November 13, 1812.

23. "The Chestnut-Hill Machine," *Aurora General Advertiser* (Philadelphia), November 27, 1812.

24. "The Perpetual Motion," *Aurora General Advertiser* (Philadelphia), November 30, 1812.

25. Thomas Jefferson to Robert Patterson, December 12 and 27, 1812, quoted in Silvio A. Bedini, *Thomas Jefferson, Statesman of Science* (New York: Macmillan, 1990), 427.

26. Robert Patterson to Thomas Jefferson, November 30, 1812, Founders Online, National Archives, http://founders.archives.gov/documents/Jefferson/03-05-02-0404-0001.

27. Arthur Ord-Hume, *Perpetual Motion: The History of an Obsession* (New York: St. Martin's, 1977), 126.

28. Henry B. Morton, "The Redheffer Perpetual Motion Machine," *Journal of the Franklin Institute* (Philadelphia) 139 (1895): 250.

29. Morton, "Redheffer Perpetual Motion Machine," 249.

30. "Perpetual Motion," *Salem (MA) Gazette*, January 5, 1813. Isaiah Lukens was a talented mechanic and clockmaker. He became a vice president of the Franklin Institute (founded in 1824) and chairman of the institute's Sciences and Arts Committee.

31. "Perpetual Motion," *Aurora General Advertiser* (Philadelphia), December 30, 1812. This article was republished in the *Democratic Republican* (Philadelphia), January 18, 1813.

32. Arthur Ord-Hume explained the deception behind Lukens's model:

The model works beautifully: if the weights are taken out of the little trucks the thing comes to rest, and the moment they are replaced it starts up again. But the apparently solid baseboard of the model is built up from pieces of thin wood and conceals a hollow centre within which is a cleverly made clockwork motor of appreciable power and slender thickness. The whole "perpetual motion machine" is covered by a glass case with four ornaments on top. One of these knobs is the winder and an attendant can keep the machine wound daily by the simple pretense of polishing the case. Motion from the clockwork motor drives a small plate on which rests the pivot of the central vertical shaft and the friction of the components is arranged so that when the weights are removed from the little trucks, the friction between pivot spindle and plate will be insufficient to transmit the continual motion for the motor, but as soon as the weights are put back, sufficient friction is restored and the plate turns the shaft.

Ord-Hume, *Perpetual Motion*, 127–28. George Escol Sellers (grandson of Charles Willson Peale) vividly remembered Redheffer's machine and Lukens's model. *Early Engineering Reminiscences (1815–40) of George Escol Sellers*, ed. Eugene S. Ferguson, *Smithsonian Institution Bulletin* 238 (Washington, DC: Smithsonian Institution, 1965), 79–86.

33. "Native Genius," *Aurora General Advertiser* (Philadelphia), January 9, 1813.

34. Robert Patterson to Thomas Jefferson, January 12, 1813, Founders Online, National Archives, http://founders.archives.gov/documents/Jefferson/03-05-02-0475. A notice in the *Aurora General Advertiser* informed readers that the broadside was for sale "at the following places, (price 25 cents)—Mr. *Wm Y Birch's*, No 31, south Second street, *J L Fernagus's* NO 98 Market street—*R. Desilver's* No 110, Walnut street—*T. Desilver's* No. 220 Market street—and at the Museum." *Aurora General Advertiser* (Philadelphia), January 13, 1813. A month earlier, Jefferson had suggested that Patterson himself should "give a popular demonstration of the insufficiency of the ostensible machinery, and of course the necessary existence of some hidden mover[.] And who could do it with more effect on the public mind than yourself?" Thomas Jefferson to Robert Patterson, December 12, 1812, quoted in Bedini, *Thomas Jefferson, Statesman of Science*, 427. Lukens's amusement included allowing Myers Fisher, a prominent Philadelphia lawyer and perpetual motion enthusiast, to make a fool of himself by reading an essay Fisher composed "in vindication of the principle of self motion. This he read with much parade to a number of us who were collected by an invitation from Lukins to examine his model." Cope, *Philadelphia Merchant*, 277.

35. *Columbian* (New York), January 6, 1813.

36. Robert J. Allison, *Stephen Decatur: American Naval Hero, 1779–1820* (Amherst: University of Massachusetts Press, 2005); Walter R. Borneman, 1812: The War That Forged a Nation (New York: HarperCollins, 2004).

37. *New York Evening Post*, January 8, 1813. The *New York Herald* printed the same article on January 9.

38. "Perpetual Motion" *Enquirer* (Richmond, VA), January 16, 1813.

39. Cadwallader D. Colden, *The Life of Robert Fulton* (New York: Kirk and Mercein, 1817), 219. The *Commercial Advertiser* embellished the scene even further. The man turning the crank was described as "a poor lank half-starved, grisly bearded wretch starting from his chair (which stood near a very warm and comfortable fire, and a convenient cupboard with crusts of bread and parings of cheese scattered over

its surface)." The man threw up his hands in supplication and "slunk sideways in the true Jerry Sneak style into one corner of the room, where he begged most piteously for mercy." *New York Commercial Advertiser*, January 11, 1813.

40. *National Advocate* (New York), January 9, 1813. Some reports and later writers claim this was Redheffer, but there is no evidence that it was. The dates are wrong: Redheffer was still in Philadelphia in January 1813.

41. *Aurora General Advertiser* (Philadelphia), January 15, 1813.

42. *Aurora General Advertiser* (Philadelphia), January 25, 1813.

43. *Journal of the Twenty Third House of Representatives of the Commonwealth of Pennsylvania* (Harrisburg: J. Peacock, printer, 1812), 285–86.

44. "Perpetual Motion!," *Tickler* (Philadelphia), March 10, 1813.

45. *Daily National Intelligencer* (Washington, DC), October 26, 1813.

46. *Northern Whig* (Hudson, NY), February 23, 1813.

47. *Newburyport (MA) Herald and Country Gazette*, November 16, 1813.

48. *Tickler* (Philadelphia), February 9, 1813.

49. *New York Columbian*, October 30, 1815; *Daily National Intelligencer* (Washington, DC), November 10, 1815. The *Salem Gazette* (November 7, 1815) announced that another perpetual motion machine, "by a Frenchman," was on exhibit in Boston.

50. "No News," *Salem (MA) Gazette*, July 19, 1816.

51. *National Standard* (Middlebury, VT), November 8, 1815.

52. "Perpetual Motion," *Boston Commercial Gazette*, July 26, 1816.

53. *Niles Weekly Register* (Baltimore), September 7, 1816.

54. In 1819 Redheffer announced plans to apply for a patent. *Massachusetts Spy* (Boston), November 24, 1819.

55. *American Beacon and Commercial Diary* (Norfolk, VA), September 5, 1816, reprint from the *American Daily Advertiser* (Philadelphia).

56. *Northern Whig* (Hudson, NY), September 17, 1816, reprint from the *Binghamton (NY) Phoenix*.

57. "Wonderful!," *Alexandria (VA) Gazette Commercial and Political*, February 8, 1817.

58. *American Journal* (Ithaca, NY), August 25, 1819.

59. *Easton (MD) Gazette and Eastern Shore Intelligencer*, March 15, 1819; *Cherry Valley (NY) Gazette*, August 17, 1819.

60. *Plough Boy* (Albany, NY), September 4, 1819. Foster then took his machine to New York City. *New York Columbian*, December 18, 1819.

61. *New York Columbian*, September 21, 1819.

62. *Poulson's American Daily Advertiser* (Philadelphia), March 4, 1817.

63. *American Daily Advertiser* (Philadelphia), September 5, 1818.

64. *Watchtower*, October 12, 1818, reprint from the *Philadelphia Gazette*; *New York Evening Post*, July 20, 1819, reprint from *Relf's Philadelphia Gazette, and Daily Advertiser*, July 19, 1819; *Massachusetts Spy* (Boston), July 21, 1819.

65. *Essex Register* (Salem, MA), March 3, 1821.

66. *American Mercury* (Hartford, CT), April 3, 1827; *Rhode Island American, Statesman and Providence Gazette*, March 27, 1827.

67. The Philadelphia physician Benjamin Rush later recalled that "many facts were circulated that tended to create a belief that the man himself did not believe in his supposed discovery and that he had contrived it only to obtain money, in which he

succeeded." *The Autobiography of Benjamin Rush: His "Travels through Life" together with His Commonplace Book for 1789–1813*, ed. George W. Corner (Princeton, NJ: Princeton University Press, 1948), 309.

68. *The Diary of William Bentley, Pastor of the East Church, Salem, Massachusetts*, vol. 4 (Salem, MA: Essex Institute, 1914), 615, entry for September 17, 1819.

69. Chandos Brown described William Bentley as the "Pepys of Salem." Chandos Brown, "A Natural History of the Gloucester Sea Serpent: Knowledge, Power and Science in Antebellum America," *American Quarterly* 42, no. 3 (September 1990): 408n9.

70. Dionysius Lardner, *Popular Lectures on the Steam Engine, in which its construction and operation are familiarly explained: with an historical sketch of its invention and progressive improvement with additions by James Renwick* (New York: printed for E. Bliss, 1828). Lardner lectured in the United States between 1840 and 1844, visiting towns and cities along the East Coast and as far west as St. Louis. See Lardner, *Popular Lectures on Science and Art: Delivered in the Principal Cities and Towns of the United States* (New York: Greeley and McElrath, 1845); A. L. Martin, *Villain of Steam: A Life of Dionysius Lardner* (Carlow, Ireland: Tyndall Scientific, 2012).

71. Brooke Hindle and Steven Lubar, *Engines of Change: The American Industrial Revolution, 1790–1860* (Washington, DC: Smithsonian Institution Press, 1986), 15–17; William Rosen, *The Most Powerful Idea in the World: A Story of Steam, Industry and Invention* (Chicago: University of Chicago Press, 2010). In 1772 Christopher Colles of Philadelphia announced a proposed course on hydrostatics, including a working model of a steam engine. "Natural Philosophy," *Pennsylvania Packet, and General Advertiser* (Philadelphia), March 23, 1772; Christopher Colles's lecture advertisement reprinted in York, *Mechanical Metamorphosis*, 51.

72. "Steam mills have not yet been adopted in America, but we shall probably see them after a short time in New-England and other places, where there are few mill seats and in this and other great towns of the United States. The city of Philadelphia, by adopting the use of them, might make a saving of above five per cent on all the grain brought hither by water, . . . and they might be usefully applied to many other valuable purposes." Tench Coxe, *An Address to the Assembly of the Friends of American Manufactures* (Philadelphia: Robert Aiken and Sons, 1787), 9.

73. Andrea Sutcliffe, *Steam: The Untold Story of America's First Great Invention* (New York: Palgrave Macmillan, 2004), 163–64.

74. York, *Mechanical Metamorphosis*, 53.

75. Sutcliffe, *Steam*, 68; Thompson Westcott, *The Life of John Fitch, the Inventor of the Steamboat* (Philadelphia: J. B. Lippincott, 1857), 251.

76. Fitch's steamboat service lasted only a few months in the summer of 1790. Sutcliffe, *Steam*, 91–94.

77. "To the Encouragers of Useful Arts," *Pennsylvania Journal* (Philadelphia), February 11, 1786.

78. Tench Coxe suggested that the federal government did have ample resources, in the form of western lands, at its disposal with which to reward invention. Coxe, *An Address to the Assembly of the Friends of American Manufactures* (Philadelphia: R. Aiken and Sons, 1787), 21.

79. Tench Coxe, *Observations on the Agriculture, Manufactures and Commerce of the United States* (New York: Francis Childs and John Swaine, 1789), 18–19. Jacob E.

Cooke, "Tench Coxe, Alexander Hamilton, and the Encouragement of American Manufactures," *William and Mary Quarterly* 32, no. 3 (July 1975): 369–92; Jacob E. Cooke, *Tench Coxe and the Early Republic* (Chapel Hill: University of North Carolina Press, 1978).

80. Article 1, Section 8, reads: "To promote the Progress of Science and useful Arts, by securing for limited Times to Authors and Inventors the exclusive Right to their respective Writings and Discoveries." The Patent Act was passed April 10, 1790: "An Act to Promote the Progress of Useful Arts." In the 1780s several states, including Pennsylvania, issued patents. But there was no consistency for what was patentable or how long a patent would last. York, *Mechanical Metamorphosis*, 192.

81. The 1790 law established an examination board consisting of the secretary of state, the secretary of war, and the attorney general. Adding the investigation of the validity of a patent application to the already heavy workloads of the chief cabinet officers quickly became impractical. Hunter Dupree, *Science in the Federal Government: A History of Policies and Activities* (Cambridge, MA: Harvard University Press, 1957), 11–13.

82. "Extract of a Letter from Augusta (Georgia)," *Norwich (CT) Packet and Country Journal*, March 26, 1790.

83. "To the Public," *Poulson's American Daily Advertiser* (Philadelphia), July 15, 1805; Oliver Evans, *The Abortion of the Young Steam Engineers Guide* (Philadelphia: Fry and Kammerer, 1805).

84. Greville Bathe and Dorothy Bathe, *Oliver Evans: A Chronicle of Early American Engineering* (Philadelphia: Historical Society of Pennsylvania, 1935; repr., New York: Arno, 1972), 109.

85. Bathe and Bathe, *Oliver Evans*, 112; Patrick N. I. Elisha [Oliver Evans], *Patent Right Oppression Exposed, or Knavery Detected* (Philadelphia: R. Folwell, 1813).

86. Bathe and Bathe, *Oliver Evans*, 207.

87. Oliver Evans, *Oliver Evans to His Counsel: Who Are Engaged in Defence of His Patent Rights, for the Improvements He Has Invented* (Philadelphia: Oliver Evans, 1817), 47; Bathe and Bathe, *Oliver Evans*, 214–16.

88. Bathe and Bathe, *Oliver Evans*, 216.

89. Evans, *Oliver Evans to His Counsel*, 45.

90. Colden, *Life of Robert Fulton*, 172.

91. Colden, 173.

92. *Republican Watch-Tower* (New York), August 25, 1807; Sutcliffe, *Steam*, 185.

93. Colden, *Life of Robert Fulton*, 181–82.

94. Sutcliffe, *Steam*, 183–84; Robert Fulton, *Torpedo War and Submarine Explosions* (New York: William Elliot, 1810), 53.

95. Fulton, *Torpedo War*, 53.

96. "The North River Steamboat of Clermont," *Republican Watch-Tower* (New York), July 15, 1808.

97. "The North-River Steam-Boat of Clermont and Mount Stevens," *Mercantile Advertiser* (New York), July 20, 1808.

98. Colden, *Life of Robert Fulton*, 4.

99. "The Steam-Boat," *Times and Weekly Adviser* (Hartford, CT), September 10, 1822.

100. Geoffrey Miller, "The Alida Waltz: Early 19th Century Music and the Hudson River," Hudson River Maritime Museum website, http://www.hrmm.org/history-blog/the-alida-waltz-early-19th-century-musicand-the-hudson-river.

101. J. C. Stoddard, Musical Instrument, Patent no. 13668, October 9, 1855.

102. "This morning a little before 7 o'clock, the President of the U. States left this city [Baltimore] in the steamboat Philadelphia, for the eastward." "Abstract of Intelligence by This Mornings Mail," *Boston Intelligencer and Morning and Evening Advertiser,* June 7, 1817.

103. "Christmas and New Year's Presents," *Boston Intelligencer and Evening Gazette,* January 9, 1819; Jacob Abbott, *Rollo's Travels* (Philadelphia: Hogan and Thompson; Boston: Gould, Kendall and Lincoln, 1839).

104. Catherine Lynn, *Wallpaper in America: From the Seventeenth Century to World War I* (New York: W. W. Norton, 1980), 172. Information on bandboxes, 292–300. The *Vues d'Amerique du nord* wallpaper from the Stoner house was rescued before the house was torn down. It was reinstalled in the White House Diplomatic Reception room in 1962.

105. "Norfolk Steam Mill," *American Beacon and Norfolk-Portsmouth (VA) Daily Advertiser,* January 3, 1820.

106. "Patent Roller Gins," *City Gazette and Daily Advertiser* (Charleston, SC), December 20, 1802.

107. "Taylor and Roebuck," *Philadelphia Gazette and Universal Daily Advertiser,* March 3, 1800.

108. "Steam Engine," *Poulson's American Daily Advertiser* (Philadelphia), September 21, 1801.

109. "Spring and Summer Goods," *Columbian Phenix or Providence (RI) Patriot,* April 20, 1811; "Shipman, Clarke and Co.," *Connecticut Journal* (Hartford), January 17, 1814; "Fall Goods," *Albany (NY) Daily Advertiser,* October 2, 1815; "Superfine London Broadcloths," *Massachusetts Spy* (Boston), January 1, 1817; "New Store," *Baltimore Patriot Evening Advertiser,* May 3, 1815.

110. "New day line between Albany & New-York" (Albany, NY, 1839), New-York Historical Society, Broadside SY1839 no. 52.

111. "For Sale," *Mercantile Advertiser* (New York), April 3, 1810.

112. "Loring & Goodman," *New-York Gazette and General Advertiser,* June 4, 1812.

113. "The Steam Boat Stages," *Baltimore Patriot and Evening Advertiser,* January 16, 1817.

114. The first locomotive built for the South Carolina Railroad was called the *Best Friend of Charleston.* Hindle and Lubar, *Engines of Change,* 127.

115. Hindle and Lubar, 126.

116. James D. Dilts, *The Great Road: The Building of the Baltimore and Ohio, the Nation's First Railroad, 1828–1853* (Stanford, CA: Stanford University Press, 1993), 46.

117. *Baltimore Patriot and Mercantile Advertiser,* January 15, 1830. "Our readers must be deeply interested in the success of the scheme [of B&O]. The prosperity of our country is intimately identified with that of Baltimore, the trade of our citizens with that port being considerable."

118. Hank Wieand Bowman, *Pioneer Railroads* (Greenwich, CT: Fawcett, 1954), 23, cited in Norm Cohen, *Long Steel Railroad: The Railroad in American Folksong* (Urbana: University of Illinois Press, 2000), 39.

119. Cohen, *Long Steel Railroad*, 39.

120. William H. Brown, *The History of the First Locomotives in America from the Original Documents, and the Testimony of Living Witnesses* (New York: D. Appleton, 1874), 119.

121. *American Railroad Journal and Advocate of Internal Improvements* (New York) 2, no. 1 (January 5, 1833).

122. "Locomotive," *Salem (MA) Gazette*, July 14, 1835.

123. *Niles Weekly Register* (Baltimore), March 17, 1927, 32, 33, quoted in Dilts, *Great Road*, 46.

124. *The Great Steam Duck; or, A Concise Description of a Most Useful and Extraordinary Invention for Aerial Navigation, Authored by an Anonymous "Member of the L.L.B.B."* (Louisville, KY: Henkle, Logan, 1841).

125. "High Pressure Poetry: Verses suggested by a conversation respecting the astonishing rate at which steam carriages are expected to go, and the consequent march of refinement," *Columbian Register* (New Haven, CT), December 14, 1833.

126. "Locomotion: Walking by Steam, Riding by Steam, Flying by Steam" (ca. 1830), Princeton University Graphic Arts Collection, https://graphicarts.princeton. edu/2016/09/25/self-walking-boots/.

127. William Heath, "March of Intellect No. 2" (1829), Princeton University Graphic Arts Collection, https://www.princeton.edu/~graphicarts/2013/07.

128. McCormick Collection of Aeronautica, item 284, no. 60, Princeton University; also in Colonial Williamsburg Collection, Costume Accessories, 1841.

129. Hermann Noordung, *The Problem of Space Travel: The Rocket Motor* (Washington, DC: National Aeronautics and Space Administration, 1995), 16.

130. "Steam," *Salem (MA) Gazette*, March 12, 1830. "Horseless and driverless carts and carriages came rattling down the highways horseless and driverless, wheelbarrows trundled along without any visible agency."

131. Lisa Hopkins, "Jane C. Loudon's *The Mummy!*: Mary Shelley Meets George Orwell, and They Go in a Balloon to Egypt," *Cardiff Corvey: Reading the Romantic Text* 10 (June 2003), accessed June 19, 2019, http://www.cf.ac.uk/encap/corvey/articles/cc10_n01.pdf.

132. *Nathaniel Hawthorne: The American Notebooks*, ed. Claude M. Simpson (Columbus: Ohio State University Press, 1972), 248–49, entry for Saturday, July 27, 1844.

133. *National Anti-Slavery Standard* (New York), January 25, 1868.

134. E. F. Bleiler, "From the Newark Steam Man to Tom Swift," *Extrapolation* 30, no. 2 (Summer 1989): 102.

135. Reprint of newspaper article in the *Bruce Herald* (Milton, New Zealand), June 3, 1868.

136. "Steam Carriage," U.S. Patent no. 75,874, issued March 24, 1868. Report of Dederick's demonstration of the Steam Man appeared in the *Daily Advertiser* (Newark, NJ), January 9, 1868, and the *Newark Observer*, January 23, 1868. Subsequently, news spread throughout the country and as far away as New Zealand. *Farmer's Cabinet* (Amherst, NH), March 26, 1868; "A Steam Man," *Bruce Herald* (Milton, New Zealand), June 3, 1868; Bleiler, "From the Newark Steam Man to Tom Swift," 101–3, 105; Lisa Nocks, *The Robot: The Life Story of a Technology* (Westport, CT: Greenwood, 2006), 50; Edward S. Ellis, *The Steam Man of the Prairies*, *Beadle's American Novel* no.

45, August 1868. Edward Ellis lived in northern New Jersey and almost certainly read the newspaper reports of Dederick's invention.

137. In the 1870s and 1880s, an entire series of Frank Reade Jr. stories featured the steam man and steam horses. Bleiler, "From the Newark Steam Man to Tom Swift," 105.

138. See chap. 6, "Exploding Steamboats and the Culture of Calamity," in Cynthia A. Kierner, *Inventing Disaster: The Culture of Calamity from the Jamestown Colony to the Johnstown Flood* (Chapel Hill: University of North Carolina Press, 2019).

139. "According to one estimate, roughly 30 percent of all steamboats built in the United States before 1849 were destroyed in explosions or other accidents." Kierner, *Inventing Disaster*, 170.

140. Kierner, 178.

141. Harriet Beecher Stowe, "The Canal Boat," *Godey's Lady's Book*, 1841.

142. He noted, "It would not have been suspected to have had a sea origin had it not been for the protuberances by which it is distinguished & for the agreement they have with the representation of the Sea Serpent." Bentley, *Diary*, 4:481, entry for September 30, 1817.

143. Bentley, *Diary*, 4:467–68, entry for July 15, 1817.

144. Fitz-Greene Halleck, *Fanny* (New York: C. Wiley, 1819), 20–21.

145. *The Diary of George Templeton Strong*, ed. Allan Nevins and Milton Halsey Thomas, vol. 3 (New York: Macmillan, 1952), 409, entry for August 10, 1858.

4. Grand Designs

1. Cotton Mather, "Advice from Taberah: A Sermon Preached after the Terrible Fire" (Boston: B. Green, 1711).

2. Jonathan Mayhew, "God's Hand and Providence to be religiously acknowledged in public Calamities: A Sermon Occasioned by the Great Fire in Boston, New England" (Boston: Richard Draper, 1760).

3. Thomas Pownall, "A Brief" (Boston: Richard Draper, 1760).

4. In 1835 fire again destroyed over thirteen acres of buildings. Gerard T. Koeppel, *Water for Gotham: A History* (Princeton, NJ: Princeton University Press, 2001), 175, 125.

5. Drinker, *Diary*, 3:1617–18, entries for January 14 and 15, 1803. There are more entries for fires in Drinker's diary than almost any other subject (second only to stomach disorders). Nelson M. Blake, *Water for the Cities: A History of the Urban Water Supply Problem in the United States* (Syracuse, NY: Syracuse University Press, 1956), 5.

6. N. Blake, *Water for the Cities*, 13, 14.

7. Benjamin Henry Latrobe, *View of the Practicality and Means of Supplying the City of Philadelphia with Wholesome Water in a Letter to John Miller, Esquire, December 29th, 1798* (Philadelphia: Zachariah Poulson Jr., 1799), 7.

8. Edward C. Carter II, ed., *The Virginia Journals of Benjamin Henry Latrobe, 1795–1798*, vol. 2 (1797–1798) (New Haven, CT: Yale University Press, 1977), 380; Carl S. Smith, *City Water, City Life: Water and the Infrastructure of Ideas in Urbanizing Philadelphia, Boston, and Chicago* (Chicago: University of Chicago Press, 2013), 15.

9. *Aurora General Advertiser* (Philadelphia), May 3, 1799; N. Blake, *Water for the Cities*, 12.

10. N. Blake, *Water for the Cities*, 9, quoting William Currie.

11. N. Blake, 6; Simon Finger, *The Contagious City: The Politics of Public Health in Early Philadelphia* (Ithaca, NY: Cornell University Press, 2012); Billy G. Smith, *Ship of Death: A Voyage that Changed the Atlantic World* (New Haven, CT: Yale University Press, 2013.

12. N. Blake, *Water for the Cities*, 93.

13. "Water also could contribute to moral reform by replacing alcohol as America's drink of preference. Many temperance advocates believed that bad water actually led drinkers to the bottle. Faced with a foul-tasting glass of water, ran the theory, even the most ardent opponent of alcohol would be tempted to turn to drink, or at the very least to offset the bad taste by adding some spirits." Michael Rawson, "The Nature of Water: Reform and the Antebellum Crusade for Municipal Water in Boston," *Environmental History* 9, no. 3 (July 2004): 420.

14. Loammi Baldwin Jr., *Report on Introducing Pure Water into the City of Boston*, 2nd ed. (Boston: Hilliard, Gray, 1835), iv; C. Smith, *City Water, City Life*, 180.

15. *Celebration of the Introduction of the Water of Cochituate Lake into the City of Boston, October 25, 1848* (Boston: J. H. Eastburn, 1848).

16. C. Smith, *City Water, City Life*, 14.

17. Latrobe, *View of the Practicality*.

18. N. Blake, *Water for the Cities*, 34.

19. N. Blake, 12.

20. Latrobe, *View of the Practicality*, 18.

21. *Philadelphia Gazette and Universal Daily Advertiser*, January 28, 1801.

22. "Temple of the Muses," *Philadelphia Repository, and Weekly Register*, April 25, 1801.

23. Benjamin Henry Latrobe, *Anniversary Oration, Pronounced before the Society of Artists of the United States* (Philadelphia: Bradford and Inskeep, 1811), 17.

24. Caroline Winterer, *Culture of Classicism: Ancient Greece and Rome in American Intellectual Life, 1780–1910* (Baltimore: Johns Hopkins University Press, 2001), 19–20.

25. William Rush, *Water Nymph with Bittern*. Rush was a member of the Watering Committee.

26. "The air produced by the agitation of water is of the purest kind, and the sudden evaporation of water, scattered through the air, absorbs astonishing quantities of heat,—or to use the common phrase, creates a great degree of cold." Latrobe, *View of the Practicality*, 18.

27. George Blake, *Blake's Collection of Duetts for Two Flutes, Clarinets, or Violins* (Philadelphia: George E. Blake, 1807). The illustration is Thomas Birch's engraving, "The Waterworks, Centre Square, Philadelphia."

28. "Centre Square Philad'a," *The Casket: Flowers of Literature, Wit and Sentiment* (Philadelphia, October 1831).

29. Elizabeth Milroy, *The Grid and the River: Philadelphia's Green Places, 1682–1876* (University Park: Penn State University Press, 2016).

30. N. Blake, *Water for the Cities*, 86–87. "Seen from across the Schuylkill, the water works suggested the cluster of temples and porticos surrounding some ancient Greek acropolis. Equally striking was the vista presented on the other side, where a railed terrace and steps formed a handsome walk along the side of the mill race and across the top of the head arches and the earthen dike out to the

charming colonnaded pavilion, from which one could look down on the water passing over the dam. Once again, as at Centre Square, the authorities had sought to serve the aesthetic needs of the city while assuaging its thirst." N. Blake, *Water for the Cities*, 87.

31. Jane Mork Gibson and Robert Wolterstorff, "The Fairmount Waterworks," *Philadelphia Museum of Art Bulletin* 84, nos. 360/361 (Summer 1988): 26–27; Henri Marceau, *William Rush: The First Native American Sculptor* (Philadelphia: Pennsylvania Museum of Art, 1937), 28; D. Dodge Thompson, "The Public Work of William Rush: A Case Study in the Origins of American Sculpture," in *William Rush, American Sculptor*, ed. Richard D. Boyle (Philadelphia: Pennsylvania Academy of Fine Arts, 1982), 41.

32. John P. Sheldon to Eliza Whiting Sheldon, December 10, 1825, in John P. Sheldon, "A Description of Philadelphia in 1825," *Pennsylvania Magazine of History and Biography* 60, no. 1 (January 1936): 74.

33. Thomas Ewbank, *A Descriptive and Historical Account of Hydraulic and Other Machines for Raising Water*, 4th ed. (New York: Greeley & McElrath, 1850), 301; Gibson and Wolterstorff, "Fairmount Waterworks," 26. Ewbank later became US commissioner of patents (1849–52).

34. *Savannah (GA) Republican*, May 11, 1827.

35. Trask, *Black Hawk*, 301.

36. Thomas Doughty's painting of the Fairmount Water Works was exhibited at the Academy of Fine Arts in 1822. An engraving of it by Chepas G. Childs was sold to subscribers. Carey and Lea used Child's engraving as the frontispiece for their guidebook, *Philadelphia in 1824*. Milroy, *Grid and the River*, 190.

37. *The Stranger's Guide to the Public Buildings, Places of Amusement, Streets, Lanes, Alleys, Roads, Avenues, Courts, Wharves, Principal Hotels, Steam-Boat Landings, Stage Offices, Etc. Etc. of the City of Philadelphia and Adjoining Districts* (Philadelphia: H. S. Tannes, 1828).

38. George Appleton, *A Handbook for the Stranger in Philadelphia* (Philadelphia: George S. Appleton, 1849), 25–26.

39. Bowen also provides details about the first steam-powered pumps and water-powered pumps, and the quantity of water raised every day. Eli Bowen, *The Pictorial Sketch-Book of Pennsylvania* (Philadelphia: William Bromwell, 1852), 19.

40. Gibson and Wolterstorff, "Fairmount Waterworks," 25.

41. Appleton, *Handbook for the Stranger*, 26.

42. Appleton, 29.

43. *Savannah (GA) Republican*, June 18, 1828.

44. Frances Trollope, *Domestic Manners of the Americans*, vol. 2 (London: Whittaker, Treacher, 1832), 74; Gibson and Wolterstorff, "Fairmount Waterworks," 28.

45. Charles Dickens, *American Notes for General Circulation* (London: Chapman and Hall, 1842; repr., New York: Penguin, 1985), 89. Cited in Gibson and Wolterstorff, "Fairmount Waterworks," 28.

46. Christopher M. Roberts, "The Water Works: A Place 'Wondrous to Behold,'" *Delaware River Basin Commission Annual Report* (West Trenton, NJ: DRBC, 1992), 29. Accessed June 26, 2016, https://www.nj.gov/drbc/about/public/annual-reports.html.

47. "Philadelphia, June 23," *Ariel: A Semimonthly Literary and Miscellaneous Gazette* 6, no. 5 (Philadelphia) (June 23, 1832): 80; Milroy, *Grid and the River*, 189.

48. Gibson and Wolterstorff, "Fairmount Waterworks," 28. "The quadrille had been introduced to America in the early 19th century and was reaching the height of its popularity when this music was published. Quadrille parties blossomed in public halls and in fashionable private homes all over the city." Bob Skiba, "Dance and the Urban Landscape in the 19th Century," *Philadelphia Dance History Journal*, January 25, 2012, accessed June 26, 2016, https://philadancehistoryjournal.wordpress.com/tag/fairmount-water-works/.

49. George Brewer, *A Description of the Mammoth Cave of Kentucky and Niagara River and Falls, and the Falls in Summer and Winter; the Prairies, or Life in the West; the Fairmount Water Works and Scenes on the Schuylkill, &c. &c. To Illustrate Brewer's Panorama* (Boston: J. M. Hewes, 1859), 4. The term *panorama* was first used in advertisements for Robert Barker's semicircular view of London displayed in 1791. Ralph Hyde, *Panoramania!* (London: Trefoil, 1988), 20.

50. Brewer, *Description of the Mammoth Cave*, 5.

51. Milroy, *Grid and the River*, 190.

52. "Again, Graff married the ancient and the modern in the pavilions, evoking both the engineering achievements of antiquity and the popular classical follies of contemporary landscape gardening. The complex as an engine of change that affirmed the resonant memories of its site. As a facility designed to deliver fresh water to the city, the waterworks became a powerful emblem of civic health and progress." Milroy, *Grid and the River*, 187.

53. "Fair Mount Near Philadelphia," Patriotic America: Blue Printed Pottery Celebrating a New Nation. Accessed April 18, 2021, http://www.americanhistoricalstaffordshire.com/pottery/printed-designs/patterns-shapes-by-maker/joseph-stubbs.

54. Gibson and Wolterstorff, "Fairmount Waterworks," 4. A dinner plate with an image of the Fairmount Water Works is on display in the waterworks museum.

55. Josiah Quincy to Joseph S. Lewis, April 20, 1825, Etting Collection, Historical Society of Pennsylvania; N. Blake, *Water for the Cities*, 160.

56. Blake notes the 1832 epidemic mortality rate in Philadelphia was "much lighter than in New York and other cities, and this partial deliverance was attributed to the liberal use of water in keeping the streets clean." N. Blake, *Water for the Cities*, 93, 133. For the New Orleans cholera, see Thomas H. O'Connor, *The Athens of America: Boston, 1825–1845* (Amherst: University of Massachusetts Press, 2006), 74; Charles E. Rosenberg, *The Cholera Years* (Chicago: University of Chicago Press, 1962); and Gerald F. Pyle, "The Diffusion of Cholera in the United Sates in the Nineteenth Century," *Geographical Analysis* 1, no. 1 (1969): 59–75.

57. The Manhattan Water Company opened its hydrants for free on a regular basis to clean the streets. And it allowed fire companies free access to the water, first by drilling directly into a street pipe, and later (in 1807) by installing hydrants. N. Blake, *Water for the Cities*, 61.

58. Thompson, "Public Work of William Rush," 43.

59. New Yorkers promoted the construction of a municipal water system as early as 1821. N. Blake, *Water for the Cities*, 109.

60. Charles King, *A Memoir of the Construction, Cost, and Capacity of the Croton Aqueduct [. . .] Together with an Account of the Civic Celebration of the Fourteenth October, 1842, on Occasion of the Completion of the Great Work* (New York: printed by C. King, 1843), 288.

61. King, *Memoir of the Construction*, 278.

62. King, 281–82.

63. King, 290.

64. *Phelps' New York City Guide; Being a Pocket Directory for Strangers and Citizens to the Prominent Objects of Interest in the Great Commercial Metropolis and Conductor to Its Environs* (New York: Ensign, Bridgman, and Fanning, 1854), 47.

65. King, *Memoir of the Construction*, 221. King compared it favorably with European fountains: "Indeed, there is scarcely any feature of the work more imposing and magnificent than the volume of water which its fountains pour out in perennial flow, and the height to which they are projected. There are, to be sure, higher jets in Europe—the highest perhaps in the world is that of Cassel, in Westphalia, which, according to modern travellers, rises from a pipe of 12 inches in diameter, to the extraordinary height of two hundred feet but it never plays much more than half an hour!"

66. Lydia Maria Child, *Flowers for Children* (n.p., 1854); Koeppel, *Water for Gotham*, 280.

67. Lewis H. von Vultee, "Croton Jubilee Quickstep," sheet music (New York: C. G. Christman, 1842).

68. Laura Vookles Hardin, "Celebrating the Aqueduct: Pastoral and Urban Ideas," in *The Old Croton Aqueduct: Rural Resources Meet Urban Needs*, ed. Jeffrey Kroessler (Yonkers, NY: Hudson River Museum of Westchester, 1992), no pagination.

69. Currier's *View of the Distributing Reservoir: On Murrays Hill,—City of New York* (New York: N. Currier, 1842). The color lithograph shows a bird's-eye view of the reservoir with visitors walking along the promenade atop the walls. Library of Congress Prints and Photographs Division.

70. Baron Dominique Vivant Denon, *Voyage dans la Basse et la Haute Égypte pendant les campagnes du général Bonaparte* (Paris: P. Didot l'aîné, 1802).

71. *City Gazette and Daily Advertiser* (Charleston, SC), August 29, 1809. Cosmoramas were scenes viewed through a small opening or window. Several of these were displayed in a single room. Hyde, *Panoramania!*, 125.

72. *New York Commercial Advertiser*, February 23, 1815. This was probably because of the Battle of the Pyramids in July 1798, where the French defeated Mamluk forces.

73. *New York Commercial Advertiser*, February 23, 1815.

74. Painted by Frederic Catherwood. Bernard Comment, *The Painted Panorama* (New York: Harry N. Abrams, 1999), 56.

75. George Robins Gliddon was the son of first US consular agent at Alexandria and the first individual to lecture on Egyptology in America. William Dinsmoor, "Early American Studies of Mediterranean Archaeology," *American Philosophical Society Proceedings* 87, no. 1 (1943): 97–98. Brewer's *Grand Moving Panorama* (1850) included views along the Nile. Peter E. Palmquist and Thomas R. Kailbourn, *Pioneer Photographers from the Mississippi to the Continental Divide: A Biographical Dictionary, 1839–1865* (Stanford, CA: Stanford University Press, 2005), 125.

76. Richard G. Carrott, "The Neo-Egyptian Style in American Architecture," *Antiques* 90, no. 4 (October 1966): 485.

77. Carrott, 485. John Haviland was well acquainted with ancient Egyptian design; he owned a complete set of the *Description de L'Égypte*.

78. Tourists were directed to take the Fifth Avenue stages, "which leave Fulton street and pass up Broadway, convey passengers the whole distance to the Reservoir, for 6 cents." *Phelps' New York City Guide*, 51. The reservoir was taken down in 1896, to make way for the New York Public Library. David Soll, *Empire of Water: An Environmental and Political History of the New York City Water Supply* (Ithaca, NY: Cornell University Press, 2013), 31.

79. King, *Memoir of the Construction*, 229.

80. King, 234.

81. King, 227.

82. King, 301.

83. Rawson, "Nature of Water," 415.

84. Rawson, 416.

85. *To All Who Want a Supply of Pure Water* (Boston, 1845), quoted in N. Blake, *Water for the Cities*, 203.

86. Fire companies with their engines were present in large numbers. The Lafayette Company No. 18 brought their dog Tiger "dressed in gala colors, and appearing as much in his element as any of his associates, with whom he has attended most fires that have occurred in the City for a number of years." *Celebration of the Introduction of the Water*, 16.

87. *Celebration of the Introduction of the Water*, 22.

88. *Celebration of the Introduction of the Water*, 15.

89. *Celebration of the Introduction of the Water*, 12.

90. *Celebration of the Introduction of the Water*, 10.

91. Rawson, "Nature of Water," 426. Advocates referred to their membership as the "Cold Water Army" and pledged to drink nothing but water. John Pierpont, a Unitarian minister in Boston and a strident advocate of temperance and other reform causes, wrote a number of celebratory poems and songs with names such as "We Sing the Praise of Water," "The Cup for Me," and "Cold Water." Rawson, "Nature of Water," 420. Unlike Philadelphia's workers, in Boston the workmen were not given alcohol.

92. *Celebration of the Introduction of the Water*, 12. Quincy gave specifics of manpower, engine power, and technologies that made the aqueduct possible: "At times like these there was no cessation of labor—night forces relieved day forces—and the five steam engines attached to pumps capable of raising 12,000,000 of gallons 10 feet high in 24 hours, were constantly employed, and at times their extreme capacity failed to free the excavation. Unaided by steam this work could not have been performed in less than five years, and the expense of constructing the conduit in quicksands, subject to inundations for that length of time, would have increased in still greater proportion." *Celebration of the Introduction of the Water*, 34.

93. *Celebration of the Introduction of the Water*, 35–36.

94. *Boston Daily Atlas*, August 21, 1846; N. Blake, *Water for the Cities*, 212.

95. *Celebration of the Introduction of the Water*, 38.

96. *Celebration of the Introduction of the Water*, 42.

97. One of the few historical treatments of the social and symbolic roles played by urban fountains is Jean-Pierre Goubert's *The Conquest of Water: The Advent of Health in the Industrial Age* (Princeton, NJ: Princeton University Press, 1989).

5. Internal Improvements

1. Samuel George Morton's evaluations of Native American skulls contributed to these theories of declension. As one of the first practitioners of American ethnography, Morton claimed that ethnic and racial distinctions were exhibited in cranial size and shape; race went hand in hand with superior, or inferior, intelligence. To prove his assertions, Morton collected the skulls of Native Americans from the far West. He was not particular about how his suppliers obtained the skulls. It is evident that many were robbed from graves. Morton published his evaluation of Native Americans in *Crania Americana*. See Ann Fabian's chapter on Morton in her book *The Skull Collectors*.

2. There are many advertisements in the American newspapers for Combe's *System of Phrenology*, available at bookstores and circulating libraries.

3. Balance was key to a healthy mind: an individual might have a large Destruction faculty, but if an equally large Veneration faculty tempered Destruction, he or she would not be murderous or violent. George Combe, *Elements of Phrenology*, 7th ed. (Edinburgh: Maclachlan and Stewart, 1850), v–vii. The theory that the skull takes its outward shape from the shape and size of the brain was not new, but Gall's application of the theory to the "reading" of a skull was his innovation. John Van Wyhe, "The Diffusion of Phrenology through Public Lecturing," *Science in the Marketplace: Nineteenth-Century Sites and Experiences*, ed. Aileen Fyfe and Bernard Lightman (Chicago: University of Chicago Press, 2007), 62.

4. Anne Felicity Woodhouse, "Nicholas Biddle in Europe, 1804–1807," *Pennsylvania Magazine of History and Biography* 103 (January 1979): 23–24. Gall was on a lecture tour in Germany. John Van Wyhe, *Phrenology and the Origins of Victorian Scientific Naturalism* (Farnham, UK: Ashgate, 2004), appendix A, "Reconstructed Itinerary of Gall's Lecture Tour," 210–11.

5. Anthony A. Walsh, "Phrenology and the Boston Medical Community in the 1830s," *Bulletin of the History of Medicine* 50 (1976): 262.

6. *A Place in My Chronicle: A New Edition of the Diary of Christopher Columbus Baldwin, 1829–1835*, ed. Jack Larkin and Carolina Sloat (Worcester, MA: American Antiquarian Society, 2010), entry for March 25, 1829.

7. Baldwin, *Diary*, entry for April 25, 1830.

8. Baldwin, *Diary*, entry for April 26, 1830. Combe's book was an overnight success on both sides of the Atlantic; it generated discussion and controversy over its claims about natural laws and its arguments for a secular society. *The Constitution of Man* was not only Combe's most famous work, it was one of the best-selling books of the nineteenth century. "Combe's fundamental argument in the *Constitution* is that mankind is subject to natural laws—physical and moral. It was ignorance or disobedience of these laws that led to bad governance, warfare, crime and personal misbehavior. Knowledge of natural laws, coupled with an understanding of personal behavior in relationship to these laws, could change individuals and societies for the better. For Combe, phrenology was the tool that individuals could use to improve themselves, and that educators, reformers, and governments could employ to improve society." Anthony A. Walsh, "George Combe: A Portrait of a Heretofore Generally Unknown Behaviorist," *Journal of the History of the Behavioral Sciences* 7, no. 3 (July 1971): 275.

9. "Meet Mr. Chester Harding the painter, and have a conversation on phrenology. He is a full believer and convert to the doctrine, and has taken the dimensions of all the most distinguished heads in the country, such as the members of the Supreme Court of the U.S. Daniel Webster's &c. The largest head is that of Judge Marshall & next is that of Mr. Webster. Mr. Harding now resides at Springfield." Baldwin, *Diary*, entry for March 12, 1831. Harding painted a posthumous portrait of Baldwin after his death in 1835 (based on a miniature of Baldwin by Sarah Goodridge). Baldwin met Spurzheim as he passed through Worcester on his way to Boston in August 1832. Baldwin noted that Spurzheim "looks like a German and very much as the engravings represent him." Baldwin, *Diary*, entry for August 20, 1832.

10. Baldwin, *Diary*, entry for March 27, 1834.

11. Baldwin, *Diary*, entry for April 17, 1834.

12. Baldwin, *Diary*, entry for May 14, 1834.

13. Baldwin, *Diary*, entry for August 7, 1834.

14. Baldwin, *Diary*, entry for May 2, 1835.

15. Robert H. Collyer, *Lights and Shadows of American Life* (Boston: Redding, 1843), 26.

16. Caldwell began lecturing on phrenology after he moved to Kentucky to become a medical professor at Transylvania University in Lexington. Robert E. Riegel, "The Introduction of Phrenology to the United States," *American Historical Review* 39, no. 1 (October 1933): 73–78.

17. Advertisement for Caldwell's Boston lectures, *Salem (MA) Gazette*, June 27, 1828.

18. Anthony A. Walsh, "The American Tour of Dr. Spurzheim," *Journal of the History of Medicine and Allied Sciences* 27, no. 2 (1972): 189.

19. Walsh, 190.

20. Walsh, 190.

21. Edward Warren, *The Life of John Collins Warren, M.D.: Compiled Chiefly from His Autobiography and Journals*, vol. 2 (Boston: Ticknor and Fields, 1860), 11–12.

22. *Daily Evening Transcript* (Boston), Thursday, September 13, 1832.

23. Walsh, "American Tour of Dr. Spurzheim," 190n22.

24. Michael Sappol, *A Traffic of Dead Bodies: Anatomy and Embodied Social Identity in Nineteenth-Century America* (Princeton, NJ: Princeton University Press, 2002), 114–17.

25. Nahum Capen, *Reminiscences of Dr. Spurzheim and George Combe* (New York: Fowler and Wells, 1881), 25–26.

26. *Philadelphia Gazette and Universal Daily Advertiser*, October 19, 1832; *Daily Evening Transcript* (Boston), October 22, 1832.

27. Spurzheim became ill during the fourteenth lecture and died three weeks later on November 30, 1832.

28. Matthew H. Kaufman, *Edinburgh Phrenological Society: A History* (Edinburgh: William Ramsay Henderson Trust, 2005), 19, 236.

29. "The brain and heart were kept in a 'fireproof safe' within a 'fireproof building'—the Mastodon Museum—at 92 Chestnut Street." Walsh, "American Tour of Dr. Spurzheim," 199n57.

30. Warren, *Life of John Collins Warren*, 2:12–13; J. D. Davies, *Phrenology: Fad and Science: A 19th-Century American Crusade* (New Haven, CT: Yale University Press, 1955), 20.

31. Prince Albert was so taken with Combe's ideas that he consulted Combe about the Prince of Wales's education. Prince Albert also requested Combe to do a phrenological examination of two of the royal daughters, princesses Victoria and Alice. David Stack, *Queen Victoria's Skull: George Combe and the Mid-Victorian Mind* (London: Hambledon Continuum, 2008), 175.

32. In addition to his visits to Philadelphia, New York, and Boston, Combe traveled west via the Erie Canal to Utica, Syracuse, and Auburn.

33. Warren, *Life of John Collins Warren*, 13.

34. Combe recalled, "Mr. Nicholas Biddle, manager of the United States' Bank, called and informed me that he had attended a course of lectures given by Dr. Gall at Carlsruhe in Germany, in 1806 or 1807, and he presented to me a skull which Dr. Spurzheim had marked for him, shewing the situations of the organs as then discovered, and which had remained in his possession ever since." George Combe, *Notes on the United States of North America, during a Phrenological Visit in 1838–9–40*, vol. 1 (Edinburgh: Maclachlan, Stewart, 1841), 304. Biddle returned to Combe a few months later for a reading. Combe assessed Biddle's cerebral development in love of appreciation and in benevolence as both "very large." George Combe to Nicholas Biddle, Report on N. B.'s cerebral development, March 15, 1839, Biddle Papers Collection no. 2039, Historical Society of Pennsylvania.

35. Combe, *Notes*, 1:xxi.

36. Orson Fowler, *Human Science or Phrenology* (Philadelphia: National, 1873), 213; Madeleine B. Stern, *Heads and Headlines: The Phrenological Fowlers* (Norman: University of Oklahoma Press, 1971), 213.

37. Fowler, *Human Science or Phrenology*, 214.

38. *Downfall of Babylon, or, The Triumph of Truth over Popery* (New York), January 7, 1837.

39. "Messrs. Fowlers' Lectures on Phrenology. The Mind Intrinsically Independent of the Body," *Downfall of Babylon*, December 24, 1836.

40. *Downfall of Babylon*, January 7, 1837.

41. *Eastern Argus* (Portland, ME), November 9, 1832.

42. *Daily Madisonian* (Washington, DC), January 11, 1842.

43. *New-Hampshire Sentinel* (Keene), February 7, 1833.

44. *Pittsfield (MA) Sun*, December 5, 1844.

45. *State Gazette* (Austin, TX), May 8, 1858.

46. Janet Rice McCoy, "Dr. R. C. Rutherford, Phrenologist and Lecturer: His Public Humiliation by Matrimony," *Northwest Ohio Quarterly* 74 (Summer/Fall 2002): 155.

47. McCoy, "Dr. R. C. Rutherford," 152.

48. Ralph Waldo Emerson, "Historic Notes on Life and Letters in New England," in *The Complete Works of Ralph Waldo Emerson*, vol. 10, ed. Edward Waldo Emerson (Boston: Houghton, Mifflin, 1904), 337.

49. David Reese Meredith, *Humbugs of New-York; Being a Remonstrance against Popular Delusion; Whether Science, Philosophy, or Religion* (New York: Taylor, 1838), 21.

50. *Milwaukee Daily Sentinel*, June 8, 1852.

51. *Public Ledger* (Philadelphia), December 31, 1841.

52. *A Phrenological Chart: Presenting a Synopsis of the Doctrine of Phrenology. Also, an Analysis of the Fundamental Powers of the Human Mind* [. . .] *Together with the*

Phrenological Character of [blank] / *Examined by Rev. Josiah M. Graves* (Hartford, CT: Hurlbut and Williams, printers, 1839).

53. J. D. L. Zender's "Phrenological chart, or else: a physiognomico-craniological delineation of the person of M _____," American Philosophical Society copy.

54. Samuel L. Clemens and Charles Neider, *The Autobiography of Mark Twain* (New York: Harper, 1959), 64–65.

55. "The Phrenological Developments of Charles Dickens . . . at the residence of Hon. John Davis," February 5, 1842, Misc. MSS, box 5, folder 1. American Antiquarian Society. The back of the page says "Copied by Thomas Chase from the original Chart in the Library of the American Antiquarian Society."

56. Lucretia Mott to Phoebe Post Willis, November 1, 1838, Phoebe Post Willis Papers, University of Rochester. My thanks to Carol Faulkner for providing a transcription of the letter.

57. *The Diary of James A. Garfield*, ed. Harry James Brown and Frederick D. Williams, vol. 1 (East Lansing: Michigan State University Press, 1967), 264, entry for July 10, 1854.

58. Garfield, *Diary*, entry for July 28, 1857. Garfield had a third reading done at the Fowlers' office by Nelson Sizer in 1864. This reading is simply numerical assessment on the chart included in the Fowlers' *New Illustrated Self Instructor*. The reading is dated May 14, 1864. Garfield's phrenology book is housed at the James A. Garfield National Historic Site in Mentor, Ohio (National Park Service). Mark Lintern, *The Garfield Observer* (blog), post for August 31, 2012, accessed July 30, 2015, https:// garfieldnps.wordpress.com/2012/08/31/phrenology-in-victorian-america.

59. Clemens and Neider, *Autobiography of Mark Twain*, 64.

60. There is a brief sarcastic mention of phrenology in his early novel, *Fanshawe*: "A phrenologist would probably have found the organ of destructiveness in strong development, just then, upon Edward's cranium; for he certainly manifested an impulse to break and destroy whatever chanced to be within his reach." Taylor Stoehr, *Hawthorne's Mad Scientists: Pseudoscience and Social Science in Nineteenth-Century Life and Letters* (Hamden, CT: Archon Books, 1978), 68.

61. "Phrenology," *American Magazine of Useful and Entertaining Knowledge* 2, no. 8 (April 1, 1836): 337. Hawthorne also reprinted extracts from Combe in the March and April issues.

62. "Phrenology," *American Magazine of Useful and Entertaining Knowledge*, April 1, 1836, 338.

63. Herman Melville, *The Confidence Man* (New York: Dix, Edwards, 1857), 190; Tyrus Hillway, "Melville's Use of Two Pseudo-Sciences," *Modern Language Notes* 64, no. 3 (March 1949): 147.

64. Herman Melville, *Moby-Dick; or, The Whale* (New York: Harper and Brothers, 1851), chap. 80; Hillway, "Melville's Use of Two Pseudo-Sciences"; Harold Aspiz, "Phrenologizing the Whale," *Nineteenth-Century Fiction* 23, no. 1 (June 1968): 18–27.

65. Alan Gribben, "Mark Twain, Phrenology and the 'Temperaments': A Study of Pseudoscientific Influence," *American Quarterly* 24, no. 1 (March 1972): 55.

66. Mark Twain, *Tom Sawyer* (Hartford, CT: American Publishing, 1876), chap. 12; Gribben, "Mark Twain," 58–59.

67. Edgar Allan Poe, "Critical Notices," *Southern Literary Messenger* 2, no. 3 (March 1836): 286; Edgar Allan Poe, *Essays and Reviews* (New York: Library of America, 1984), 329.

68. Poe, "Critical Notices," 286.

69. Poe, "The Murders in the Rue Morgue," *Graham's Magazine* 18, no. 4 (April 1841): 166.

70. Poe, "Murders in the Rue Morgue," 168.

71. James Kirke Paulding, *The Merry Tales of the Three Wise Men of Gotham* (New York: G. and C. Carvill, 1826), 261.

72. Paulding, 312.

73. Paulding, 318–20.

74. *The Phrenologist* was performed in Philadelphia on January 31 and February 1, 1823. Reese D. James, *Old Drury of Philadelphia: A History of the Philadelphia Stage* (New York: Greenwood, 1968), 380; *Relf's Philadelphia Gazette, and Daily Advertiser,* January 14, 1823. The play was a translation of August von Kotzebue's *Die Organe des Gehirns* (1806). The first existing English version is *The Organs of the Brain, a Comedy in Three Acts,* translated from the German by Lieut. Col. Capadose (London: Edward Bull, 1838).

75. August von Kotzebue, *The Organs of the Brain, a Comedy in Three Acts,* trans. Lieut. Col. Capadose (London: Edward Bull, 1838), 30 (act 2, scene 1).

76. Journal of J. Warner Erwin. A transcription of the diary is at https://web.archive.org/web/20060425173502/http://www.brynmawr.edu/iconog/jwe/jweint.html.

77. *Hudson River Chronicle* (Sing Sing), October 31, 1848.

78. John Neal, "The Young Phrenologist," in *New England Galaxy and United States Literary Advertiser* 18, no. 40 (October 3, 1835): 1; reprinted in *The Token and Atlantic Souvenir. A Christmas and New Year's Present,* ed. Samuel Goodrich (Boston: Charles Bowen, 1836), 156–69.

79. "January First, A.D. 3000," *Harper's New Monthly Magazine,* January 1856, 152.

80. J. G. Spurzheim, *Phrenology in Connexion with the Study of Physiognomy* (Boston: Marsh, Capen and Lyon, 1833), 109.

81. Stephen Tomlinson, "Phrenology, Education and the Politics of Human Nature: The Thought and Influence of George Combe," *History of Education* 26, no. 1 (March 1997): 2. See Tomlinson's *Head Masters: Phrenology, Secular Education, and Nineteenth-Century Social Thought* (Tuscaloosa: University of Alabama Press, 2005) for a history of Mann's friendship with George Combe and Mann's application of phrenology to education. Combe dedicated *Lectures on Popular Education* to Mann and Mann named one of his sons George Combe Mann.

82. "Phrenology Made Easy," *Knickerbocker Magazine,* June 2, 1838.

83. Nelson Sizer, *What to Do, and Why; and How to Educate Each Man for His Proper Work* (New York: Mason, Baker and Pratt, 1872). A separate section, Occupations for Women, was headed by Teaching, but also included Watchmaking and Dentistry. Sizer dedicated his useful book to Henry Ward Beecher, "the earnest defender of progressive thought," who championed phrenology when it was "everywhere spoken against."

84. "From Washington," *Southern Patriot* (Charleston, SC), February 3, 1824.

85. "To the Great Expunger," *Jamestown (NY) Journal,* March 18, 1840.

86. "A new Philosophy has sprung up within a few years past, called Phrenology. There is I believe, something in it, but not quite as much as its ardent followers proclaim. According to its doctrines, the leading passion, propensity, and characteristics of every man are developed in his head. Gall and Spurzheim, its founders or most eminent propagators, being dead, I regret that neither of them can examine the head of our illustrious Chief Magistrate. But if it could be surveyed by Dr. Caldwell, of Transylvania University, I am persuaded that he would find the organ of destructiveness prominently developed." "Debate on the President's Protest. Speech of Mr. Clay," *Baltimore Patriot and Mercantile Advertiser*, May 9, 1834.

87. Johnston included a self-portrait in *Scraps* no. 7. Seated at a table, he has a phrenology bust and books by Gall, Spurzheim, and Combe.

88. Van Buren was secretary of state and then vice president under Andrew Jackson before becoming president in March 1837. Van Buren was Jackson's handpicked successor. Johnston's cartoons were probably done during or just after the presidential election in the fall of 1836.

89. "Phrenological Measurements," *Newburyport (MA) Herald*, April 19, 1836.

90. "Phrenological Measurements."

91. Harriet Martineau, *Retrospect of Western Travel*, vol. 2 (London: Saunders and Otley, 1838), 187–88.

92. Jennifer L. Morgan, *Laboring Women: Reproduction and Gender in New World Slavery* (Philadelphia: University of Pennsylvania Press, 2004), 47.

93. Samuel Stanhope Smith, *An Essay on the Causes of the Variety of Complexion and Figure in the Human Species* (New Brunswick, NJ: J. Simpson, L. Deare, printer, 1810), 169–71.

94. George R. Price and James Brewer Stewart, *To Heal the Scourge of Prejudice: The Life and Writings of Hosea Easton* (Amherst: University of Massachusetts Press, 1999), 17–18.

95. Price and Stewart, 67.

96. For a discussion of environmentalist theory in the early nineteenth century, see Dain, *Hideous Monster*, 178–96. Phrenologists were, however, convinced that individuals, within the scope of their inherent traits—such as race and sex—had the capacity to develop under favorable circumstances. One of the most popular aspects of phrenology in the early nineteenth century was the emphasis on educating children under the principles of phrenology in order to shape their development. One of the earliest American publications on this topic is Joseph A. Warne, *Phrenology in the Family: or, The Utility of Phrenology in Early Domestic Education* (Philadelphia: George W. Donohue, 1839).

97. Charles Caldwell, *Elements of Phrenology* (Lexington, KY: A. G. Meriwether, printer, 1827), 245, 253. Caldwell was an unapologetic slave owner. Caldwell considered himself to be benevolent: "We confidently assure our flagrant philanthropist, that we are ourselves no inhuman traffickers either in 'human nerves and muscles' or in human feelings—We never purchased a slave with a view of selling him again—we have been instrumental in manumitting some, and have educated one to the profession of medicine—We are even charged by our neighbours and friends with doing an injury to our slaves by two [sic] much indulgence—Perhaps we have shown as much sympathy for the African race, and, according to our humble means, rendered them as many services, as Dr. Good has done." *Elements of Phrenology*, 260.

98. George Combe, *A System of Phrenology*, 3rd ed. (London: Longman, 1830), 600, 601, 617.

99. George Combe, "Chapter II: The Characteristics of Races, Masses, and Nations, in Part Hereditary, Section I. The Colored Race," *American Phrenological Journal and Miscellany*, September 1, 1843: 417–32.

100. Combe, *System of Phrenology*, 618. George's brother, the physician Andrew Combe, concurred: "De facto the Negro brain is inferior in intellectual power to that of the European." Andrew Combe, "Remarks on Tiedemann's Comparison of the Negro Brain and Intellect with Those of the European," *Eclectic Journal of Medicine* 2, no. 9 (July 1838): 325–28.

101. "On the American Scheme of establishing Colonies of Free Negro Emigrants on the Coast of Africa, as exemplified in Liberia," *Annals of Phrenology* (October 1, 1833): 124.

102. "Anti-Slavery Speech of Mr. Slade of Vermont, on the Abolition of Slavery and the Slave Trade in the District of Columbia. Delivered in the House of Representatives of the U.S. December 20, 1837," *Union Herald* (Cazenovia, NY), May 18, 1838.

103. This was Richard Colfax's argument in *Evidence against the Views of the Abolitionists, Consisting of Physical and Moral Proofs of the Natural Inferiority of the Negroes* (New York: James T. M. Bleakley, 1833). Colfax and other essentialists embraced a "racial modernity"—the belief in an inherent racial hierarchy and racial inequality. James Brewer Stewart defines the term in the following way: "'Racial modernity' refers to the developments I see as common to both periods—a reflexive disposition on the part of an overwhelming number of northern whites (intellectuals and politicians as well as ordinary people) to regard superior and inferior races as uniform, biologically determined, self-evident, naturalized, immutable 'truths'—and, the development of integrated trans-regional systems of intellectual endeavor, popular culture, politics and state power that enforced uniform white supremacist norms as 'self-evident' social 'facts.'" "The Emergence of Racial Modernity and the Rise of the White North, 1790–1840," *Journal of the Early Republic* 18, no. 2 (Summer 1998): 183n2.

104. David R. Roediger argues that the white working-class identity was predicated on a separation from, and debasement of, Black workers, thus contributing substantially to an increasingly defined color line in the early nineteenth century. *The Wages of Whiteness: Race and the Making of the American Working Class* (London: Verso, 1991).

105. Parades evoked scorn and criticism of Blacks performing white citizenship in a series of "bobalition" broadsides printed from the 1810s to the 1830s. See Douglas A. Jones Jr., *The Captive Stage: Performance and the Proslavery Imagination of the Antebellum North* (Ann Arbor: University of Michigan Press, 2014), 40–49.

106. Price and Stewart, *To Heal the Scourge of Prejudice*, 19.

107. Joan Pope Melish argues that the presence of Blacks in white space propelled the violence in the North: "The fact that these violent acts predated anti-abolitionist actions, persisted in concert with them, and continued after they had subsided supports the argument that the root cause of whites' anger was the presence of the 'free Negro,' and that abolitionists and printers of antislavery tracts were attacked principally because they were seen as advocates of the continuation and even growth

of that presence." Joanne Pope Melish, *Disowning Slavery: Gradual Emancipation and "Race" in New England, 1780–1860* (Ithaca, NY: Cornell University Press, 1998), 201. For the pushback against colonization, see Ousmane K. Power-Greene, *Against Wind and Tide: The African American Struggle against the Colonization Movement* (New York: NYU Press, 2014).

108. Dain, *Hideous Monster*, 177; Price and Stewart, *To Heal the Scourge of Prejudice*, 18–19.

109. Melish, *Disowning Slavery*, 204–7; Price and Stewart, *To Heal the Scourge of Prejudice*, 19–22.

110. The Fowlers recommended that "every school district should possess copies of this collection." Orson Fowler and Lorenzo Fowler, "A List of Specimens Designed for Phrenological Societies," *American Phrenological Journal and Miscellany*, April 1, 1848: 129. Madeleine Stern claims that the Fowlers were abolitionists. In the 1850s they promoted a plan for an antislavery commune in Kansas. But she also writes that the brothers deliberately kept their publications nonpartisan. *Heads and Headlines*, 173.

111. "Article V. Character of Eustache," *American Phrenological Journal and Miscellany* 2, no. 4 (January 1, 1840): 177–82. This desire for compliant Blacks was also articulated in the fiction and theater of the era. For example, George Lippard's *Blanche of Brandywine; or, September the Eleventh, 1777* (Philadelphia: G. B. Zieber, 1846) was adapted for the stage by James Gilbert Burnett and performed in New York City in 1858. The plot centers on the military conflicts of the American Revolution. Sampson, a slave, fights valiantly for the patriots. But in upholding the cause of freedom for his white master, he does not ask for his own liberty. Sampson is the loyal slave who does not overstep racial or political boundaries. Douglas A. Jones Jr. uses this play to illustrate the tacit proslavery ideology of the North in the antebellum era. Jones, *Captive Stage*, 4–5. The most iconic example of this "romantic racialism" is the slave Tom in Harriet Beecher Stowe's *Uncle Tom's Cabin; or, Life among the Lowly* (Boston: John P. Jewett, 1852).

112. Combe, *Notes*, 2:78. Lydia Maria Child expressed a similar confidence that the races could "harmonise." After viewing the casts of Native American heads at Barnum's American Museum in 1843, Child noted that "the races of mankind are different, . . . The facial angle and shape of the head is various in races and nations." She was confident that education could change, and improve, both Africans and Native Americans: "Similar influences brought to bear on the Indians or the Africans, as a race, will gradually change the structure of their skulls, and enlarge their perceptions of moral and intellectual truth. The same influences cannot be brought to bear upon them; for their past is not our past, and, of course never can be. But let ours mingle with theirs, and you will find the result varied, without inferiority. They will be flutes on different notes, and so harmonise the better." Letter 36, March 1843. Lydia Maria Child, *Letters from New York* (New York: Charles S. Francis, 1843), 251.

113. "Our apartments at the Marshall House are under the charge of a coloured man, who, although a complete negro, has a brain that would do no discredit to an European." Combe, *Notes*, 2:48–49.

114. E. B. Olmstead, "George, A Slave, Murderer of Mrs. Foster, with a View of His Skull," *American Phrenological Journal and Miscellany* (November 1, 1849): 341; Lorenzo Fowler, "Slavery Preferred to Freedom," *American Phrenological Journal and Miscellany*, November 1, 1850: 359.

115. "In promulgating the stereotype of the happy and contented bondsman, Southerners were doing more than simply putting out propaganda to counter the abolitionist image of the wretched slave. They were also seeking to put to rest their own nagging fears of slave rebellion." George M. Fredrickson, *The Black Image in the White Mind: The Debate on Afro-American Character and Destiny, 1817–1914* (Middletown, CT: Wesleyan University Press, 1971), 52.

116. The most recent account of the *Amistad* is Marcus Rediker's *The Amistad Rebellion: An Atlantic Odyssey of Slavery and Freedom* (New York: Viking, 2012). See also Howard Jones, Mutiny on the Amistad: *The Saga of a Slave Revolt and Its Impact on American Abolition, Law, and Diplomacy* (New York: Oxford University Press, 1987).

117. Marcus Rediker's analysis of the detailed playbill for *The Black Schooner* suggests that the tone of the play may have been sympathetic to the captives. Rediker, Amistad *Rebellion*, 114, 132; "The Negroes of the Amistad," *New-Hampshire Sentinel* (Keene), October 2, 1839. Rediker points out that these acrobatic displays were part of Mende culture—proving leadership and high status in society. Amistad *Rebellion*, 134.

118. George Combe, *Notes*, 3:75–77, diary entry for September 24, 1839. Combe lectured in the United States between October 1838 and June 1840. Combe also did a phrenology reading of John Quincy Adams before Adams took up the defense of *Amistad* captives before the Supreme Court. Combe's assessment of the former vice president was less than flattering. Combe found Adams to be "a man of impulse rather than of clear, sound, and consistent judgment." Adams also had "a limited intellectual capacity to perceive fine and distinct relations, combined with a self-confidence which will rarely allow him to doubt the soundness of his own inductions." Combe, *Notes*, 3:107–8.

119. Quoted in Stack, *Queen Victoria's Skull*, 224.

120. A. Cameron Grant, "George Combe and American Slavery," *Journal of Negro History* 45, no. 4 (October 1960): 262.

121. L. N. Fowler, "Phrenological Developments of Joseph Cinquez, Alias Ginqua," *American Phrenological Journal and Miscellany* 2, no. 3 (December 1839): 136–38. Simultaneous with the December edition of the *American Phrenological Journal*, the Fowlers published Cinque's image in *The Phrenological Almanac for 1840*. The Fowlers' almanacs reprinted the most interesting or unusual profiles from the previous year's *Journal*. The 1840 almanac included Black Hawk, Aaron Burr, Maria Monk, Eustache, Jinqua "leader of the Africans on board Schooner Amistad," and Antoine Le Blanc "murderer of Judge Sayre, Morristown, N.J." Fowler's report may also be the one credited to J. Fletcher, published in the *New Haven (CT) Daily Herald* on September 5. Fletcher's assessment of Cinque is very much like Fowlers': "The head is well formed and such as a phrenologist admires. In fact, such an African head is seldom to be seen, and doubtless in other circumstances would have been an honor to his race." "Phrenological Examination of Joseph Cinques, the Leader of the Revolt on board the *Amistad*," *New Haven (CT) Daily Herald*, September 5, 1839. See Richard J. Powell, "Cinqué: Antislavery Portraiture and Patronage in Jacksonian America," *American Art* 11, no. 3 (Autumn 1997): 72n10.

122. "Jinqua, the leader of the captured Africans on board the schr. *Amistad*," *New Hampshire Patriot and State Gazette* (Concord), November 21, 1840.

123. John Warner Barber, *A History of the* Amistad *Captives* (New Haven, CT: E. L. and J. W. Barber, 1840). Barber's book also contained an engraving titled *The Death of Capt. Ferrer*. Rediker notes that Barber's engraving used "racialized tropes of savagery." Rediker, Amistad *Rebellion*, 162. Moultrap's wax figures were displayed in New York at Peale's Museum and Portrait Gallery, then in Boston and Norwich, CT. In 1847 P. T. Barnum exhibited them at his museum in New York. See Rediker, Amistad *Rebellion*, 159–67.

124. Powell, "Cinqué," 54. "Cinque's image contradicted the prevailing perception of the captive Africans as savages. Instead, Cinque's portrait presented him as the embodiment of a republican (read abolitionist) ideal, an allegorical representation of Christian proselytizing and missionary work in Africa, and a symbol of black resistance and activism in the face of increasing white-on-black violence and sociopolitical unrest." Powell, "Cinqué," 63. Purvis commissioned John Sartain to make an engraving of the painting. Lithographs were sold for two dollars, which was donated to the Pennsylvania Antislavery Association. Rediker, Amistad *Rebellion*, 174.

125. Powell, "Cinqué," 57; Davison, "E. W. Clay." There may have been at least one itinerant Black phrenologist in the 1830s. "Our Correspondence Letter from Communipaw New York, Jan. 12, 1859" recalled his financially successful visit to Philadelphia in 1837. *Frederick Douglass' Paper* (Rochester, NY), January 21, 1859.

126. Rediker, Amistad *Rebellion*, 196.

127. Orson Fowler, "Sarah Kinson, or Margru," *American Phrenological Journal* 12, no. 1 (July 1, 1850): 230; Marlene D. Merrill, "Sarah Margru Kinson: The Two Worlds of an *Amistad* Captive," Electronic Oberlin Group website: http://www.oberlin.edu/external/EOG/Kinson/Kinson.html.

128. Lorenzo Fowler, "A Shelf in Our Cabinet—No. 2," *American Phrenological Journal* 35, no. 3 (March 1862): 52.

6. Fair America

1. American Philosophical Society, preface to *Transactions* (1771), 1:xvii; John C. Greene, "Science, Learning, and Utility: Patterns of Organization in the Early American Republic," in Oleson and Brown, *Pursuit of Knowledge*, 2.

2. American Academy of Arts and Sciences, preface to *Memoirs*, 1:viii.

3. American Academy of Arts and Sciences, "Charter of Incorporation, May 4, 1780," in American Academy of Arts and Sciences, *Memoirs*, 1:vii.

4. American Academy of Arts and Sciences, *Memoirs*, 1:xi.

5. Joseph Greenleaf, "An Account of an Experiment for raising Indian Corn, in poor Land," *Memoirs of the American Academy of Arts and Sciences* 1:383–85, and Benjamin Gale, "Observations on the Culture of Smyrna Wheat," *Memoirs of the American Academy of Arts and Sciences* 1:381–82.

6. Hindle, *Pursuit of Science in Revolutionary America*, 266.

7. Hindle, 275; James E. McClellan, *Science Reorganized: Scientific Societies in the Eighteenth Century* (New York: Columbia University Press, 1985), 150.

8. Edward Everett, *A Memoir of Mr. John Lowell, Junior* (Boston: Little, 1840), 65. Lowell made his will in 1832, four years before his death. Margaret W. Rossiter, "Benjamin Silliman and the Lowell Institute: The Popularization of Science in Nineteenth-Century America," *New England Quarterly* 44, no. 4 (December 1971): 606.

9. The Lowell Lectures typically paid two thousand dollars for a lecture series—approximately three times what London societies could pay. Rossiter, "Benjamin Silliman," 608.

10. *The Diary of Philip Hone, 1828–1851*, ed. Bayard Tuckerman, vol. 2 (New York: Dodd, Mead, 1889), 95, entry for November 5, 1841; Donald M. Scott, "The Popular Lecture and the Creation of a Public in Mid-Nineteenth-Century America," *Journal of American History* 66, no. 4 (March 1980): 791–809.

11. Carl Bode, *The American Lyceum* (New York: Oxford University Press, 1956); Angela Ray defines the lyceum in the nineteenth-century United States as "a discontinuous, culture-making rhetorical practice." Angela G. Ray, *The Lyceum and Public Culture in the Nineteenth-Century United States* (East Lansing: Michigan State University Press, 2005), 2.

12. Van Wyck Brooks, *The Flowering of New England* (New York: Dutton, 1936), 173–74; Ray, *Lyceum and Public Culture*, 34.

13. Brooke Hindle, "The Underside of the Learned Society in New York, 1754–1854," in Oleson and Brown, *Pursuit of Knowledge*, 102; Charter of the New York Mechanic and Scientific Institution, quoted in Hindle, "Underside of the Learned Society in New York, 1754–1854," 103.

14. Henry D. Shapiro, "The Western Academy of Natural Sciences of Cincinnati and the Structure of Science in the Ohio Valley, 1810–1850," in Oleson and Brown, *Pursuit of Knowledge*, 228.

15. Elkanah Watson, *History of the Rise, Progress, and Existing Conditions of the Western Canals in the State of New York* [. . .] *Together with the Rise, Progress, and Existing State of Modern Agricultural Societies, on the Berkshire System, from 1807, to the Establishment of the Board of Agriculture in the State of New-York, January 10, 1820* (Albany, NY: D. Steele, 1820).

16. *Poulson's American Daily Advertiser* (Philadelphia), June 19, 1822; Simon Baatz, *"Venerate the Plough": A History of the Philadelphia Society for Promoting Agriculture, 1785–1985* (Philadelphia: Philadelphia Society for Promoting Agriculture, 1985), 40. Emily Pawley explores the drive for national improvement by agricultural societies and the fairs they sponsored in *The Nature of the Future: Agriculture, Science, and Capitalism in the Antebellum North* (Chicago: University of Chicago Press, 2020).

17. "Extract from First Annual Report of the Pennsylvania Society for the promotion of Internal Improvements in the Commonwealth," *Franklin Journal and American Mechanics Magazine* 1, no. 1 (January 1826): 11.

18. "Observations on the Rise and Progress of the Franklin Institute," *Franklin Journal and American Mechanics Magazine* 1, no. 2 (February 1826): 66–70; Mission statement of the Franklin Institute of the State of Pennsylvania for the Promotion of the Mechanic Arts in the *First Annual Report of the Proceedings of the Franklin Institute of the State of Pennsylvania, for the Promotion of the Mechanic Arts* (Philadelphia: J. Harding, printers, 1825), 4–63. Bruce Sinclair, "Science, Technology, and the Franklin Institute," in Oleson and Brown, *Pursuit of Knowledge*, 196–97.

19. *Franklin Journal and American Mechanics Magazine* 1, no. 2 (February 1826): 1.

20. Thomas P. Jones (editor), "Address," *Franklin Journal and American Mechanics Magazine* 1, no. 1 (January 1826): 2.

21. Jones, 2.

22. Thomas P. Jones (editor), "Notice," *Franklin Journal* 1, no. 2 (February 1826): 64.

23. In comparison, the primary function of Benjamin Silliman's *American Journal of Science*, begun in 1818, was "the dissemination of scientific knowledge; under Silliman's editorship it also served as a bulletin board for the exchange of information." Simon Baatz, "'Squinting at Silliman': Scientific Periodicals in the Early American Republic, 1810–1833," *Isis* 82, no. 2 (June 1991): 235.

24. "Report on instruction from Board of Managers Eighth Quarterly Report," *Franklin Journal and American Mechanics Magazine* 1, no. 1 (January 1826): 3–4.

25. The Franklin Institute was originally located at Carpenter's Hall until it moved to its newly constructed building a few blocks away, on Seventh Street between Market and Chestnut, in 1826.

26. Bruce Sinclair, *Philadelphia's Philosopher Mechanics: A History of the Franklin Institute, 1824–1865* (Baltimore: Johns Hopkins University Press, 1974), 94; "Abstract of the Report of the Committee on Premiums and Exhibition on the Subject of the Third Annual Exhibition, and of the Premiums Awarded," "Quarterly Meeting of the Franklin Institute," *Franklin Journal and American Mechanics Magazine* 2, no. 5 (November 1826): 268.

27. Forty thousand people attended the 1838 exhibition. Bruce Sinclair, *Philadelphia's Philosopher Mechanics*, 94, 99; Appleton, *Handbook for the Stranger*, 41.

28. "Proposals of the Franklin Institute, for the Exhibition of October 1826, addressed to the Manufacturers of the United States," *Franklin Journal and American Mechanics Magazine* 1, no 1 (January 1826): 5.

29. "Proposals of the Franklin Institute," 5.

30. "Proposals of the Franklin Institute," 5.

31. "Proposals of the Franklin Institute," 5–10.

32. "Proposals of the Franklin Institute," 9.

33. "Abstract of the Report of the Committee on Premiums and Exhibition," 266.

34. "Proposals of the Franklin Institute," 9–10.

35. "Proposals of the Franklin Institute," 7–8.

36. *Address of the Committee on Premiums and Exhibitions of the Franklin Institute* (Philadelphia: J. Harding, 1832), 3.

37. "Abstract of the Report of the Committee on Premiums and Exhibition," 267.

38. "The aquatint, of Doughty's picture of Fair Mount Water-works," "Abstract of the Report of the Committee on Premiums and Exhibition," 267.

39. *Niles Weekly Register* (Baltimore), November 14, 1846, 170, 172. Reprinted from Solomon R. Roberts, *The Promotion of the Mechanic Arts in America: An Address Delivered at the Close of the Sixteenth Exhibition of American Manufactures, Held in Philadelphia, by the Franklin Institute of the State of Pennsylvania for the Promotion of the Mechanic Arts, October 30, 1846* (Philadelphia: printed by John C. Clark, 1846).

40. James M'Henry, *An Ode: Written by Request, on the Opening of the Exhibition at the Franklin Institute of Philadelphia, October 1828* (Philadelphia: n.p., 1828).

41. "Many protectionists, particularly those at the Franklin and later the American Institute, saw their efforts to implement a tariff as part of a greater struggle to empower the nation." Joanna Cohen, *Luxurious Citizens: The Politics of Consumption in Nineteenth-Century America* (Philadelphia: University of Pennsylvania Press, 2017), 127.

42. *Report of the Third Annual Fair of the American Institute of the City of New York Held at Masonic Hall, October, 1830* [. . .]. (New York: J. Seymour, 1830), 1–2.

43. *Report of the Third Annual Fair*, 20.

44. *Report of the Third Annual Fair*, 6.

45. *Premiums Awarded by the Managers of the Twenty-Sixth Annual Fair of the American Institute, 351 Broadway, October, 1853* (New York: W. H. Tinson, 1854).

46. *Report of the Third Annual Fair*, 9–10.

47. *Report of the Third Annual Fair*, 22.

48. *Report of the Third Annual Fair*, 22.

49. *Report of the Third Annual Fair*, 6.

50. *Report of the Third Annual Fair*, 6.

51. "Regular Toasts," *Journal of the American Institute of the City of New York* 3, no. 1 (1838): 18.

52. "Regular Toasts," 18.

53. "An Ode, Sung by Capt. Joseph Cowdin, at the Supper of the Tenth Annual Fair of the American Institute. Composed by Rufus Dawes, for the Occasion," *Journal of the American Institute of the City of New York* 3, no. 1 (1838): 24.

54. "Regular Toasts," 19.

55. "Regular Toasts," 18.

56. Edward Everett, *Address Delivered before the American Institute of the City of New York at the Fourth Annual Fair, October 14, 1831* (New York: Van Norden and Mason, 1831), 8.

57. Tristram Burges, *Address of the Hon. Tristram Burges* [of Rhode Island], *Delivered at the Third Annual Fair of the American Institute of the City of New York. Held at Masonic Hall, October, 1830* (New York: John M. Danforth, 1830), 32.

58. "Fair of the Franklin Institute of Pennsylvania," *Journal of the American Institute* 4, no. 2 (November 1838): 60–66.

59. Jeffrey A. Auerbach, *The Great Exhibition of 1851: A Nation on Display* (New Haven, CT: Yale University Press, 1999).

60. *Official Catalog of the Great Exhibition of the Works of Industry of All Nations, 1851* (London: Spicer Brothers, 1851). The United States section begins on page 312.

61. "Hereabouts the visitor may pause to notice the meat-biscuit exhibited by Borden. This is hard bread, made of flour thoroughly saturated with extract of beef. It is pronounced by competent judges to be highly nutritious, and took a council medal at London, in 1851." William Richards, *A Day in the Crystal Palace and How to Gain the Most of It* (New York: G. P. Putnam, 1853), 32–33.

62. Charles T. Rodgers, "Last Appendix to 'Yankee Doodle,'" in *American Superiority at the World's Fair* (Philadelphia: J. J. Hawkins, 1852), 91.

63. Robert F. Dalzell, *American Participation at the Great Exhibition of 1851* (Amherst, MA: Amherst College Press, 1960), 52–53. Borden's biscuit was a commercial failure. He had much greater success with his later product, condensed milk. Emily Pawley describes Cyrus McCormick's triumph at the Great Exhibition reaping trials, in *Nature of the Future*, 84.

64. "Every lover of liberty and liberal institutions should thank England for the opportunity she has afforded the world of learning this important lesson." *Springfield (MA) Republican*, September 19, 1851; Dalzell, *American Participation*, 54.

65. Rodgers, *American Superiority*, 6.

66. Rodgers, 13. Rodgers reprinted all the good press notes about American exhibits, yet the book's title, *American Superiority at the World's Fair*, betrays the sense of injustice, or inferiority, many Americans felt.

67. Horace Greeley, *Glances at Europe in a Series of Letters* (New York: Dewitt and Davenport, 1852), vi.

68. *Cleveland Plain Dealer*, December 1, 1851; Dalzell, *American Participation*, 54.

69. *New York Daily Times*, December 1, 1851.

70. *Cleveland Plain Dealer*, April 21, 1851, quoted in Dalzell, *American Participation*, 64; Edwin G. Burrows, *The Finest Building in America: The New York Crystal Palace, 1853–1858* (New York: Oxford University Press, 2018).

71. "New-York Crystal Palace for the Exhibition of the Industry of All Nations," *Plough, the Loom, and the Anvil* 5, no. 5 (November 1852): 315.

72. Samuel G. Goodrich, "The Crystal Palace in New York," *Merry's Museum and Parley's Magazine*, January 1853, 63.

73. Abel Stevens, "The New York Crystal Palace," *National Magazine: Devoted to Literature, Art, and Religion* (New York) 2, no. 1 (January 1853): 80–81.

74. Horace Greeley, *Art and Industry as Represented at the New York Crystal Palace* (New York: Putnam, 1853), 25.

75. *Wilmington (NC) Journal*, September 2, 1853.

76. Another statue of Webster and one of Washington were sculpted from spermaceti (a waxy substance from the whale). Richards' guidebook humorously commented: "The north-east corner of the court affords further examples of the fine arts. Here are busts of Washington and of Webster in spermaceti, with backgrounds of candles! Is not this rather making light of sculpture?" Richards, *Day in the Crystal Palace*, 34–35.

77. Ivan D. Steen, "America's First World's Fair: The Exhibition of the Industry of All Nations at New York's Crystal Palace, 1853–1854," *New-York Historical Society Quarterly* 47, no. 3 (July 1963): 274.

78. Earle Edson Coleman, "The Exhibition in the Palace: A Bibliographic Essay," *Bulletin of the New York Public Library* 64, no. 9 (September 1960): 459–78.

79. Richards, *Day in the Crystal Palace*, 5.

80. Richards, 9.

81. Richards, 160–61.

82. Richards, 111–12, 115.

83. "Steam power and space will be gratuitously furnished for the most interesting processes in art and industry, and as inventors and exhibitors will be permitted, under certain judicious regulations, to run the machinery for their own benefit, this branch of the Exhibition is expected to become especially interesting." Orson Desaix Munn, "The Crystal Palace Re-organized," *Scientific American*, April 19, 1854, 261.

84. Richards, *Day in the Crystal Palace*, 159.

85. Richards, 157.

86. "The very image of the Crystal Palace became an icon of conspicuous consumption, and it lent status to merchants across the United States who exhibited there. John F. Ellis, a merchant in Washington D.C., incorporated an engraving of the Crystal Palace in an advertisement for 'Stationery, Music, Pianos, Fancy Goods &c.'" Margaret Frick, "From Palace to Parlor: Exhibition Display, Consumerism, and Cultivation of Taste at the New York Crystal Palace," in *New York Crystal Palace 1853*, digital publication based on a 2017 Focus Gallery exhibition at Bard Graduate Center, accessed May 1, 2018, http://crystalpalace.visualizingnyc.org/digital-publication; *Barre (MA) Patriot*, April 7, 1854; *Farmer's Cabinet* (Amherst, NH), November 9, 1854; *Farmer's Cabinet*, March 1, 1866; *Washington (DC) Sentinel*,

June 1, 1854. The same issue contained Barnum's detailed advertisement about the Crystal Palace reopening. *New York Herald*, January 9, 1854; *Harper's Weekly* (New York), December 26, 1857.

87. Richards, *Day in the Crystal Palace*, 163–64. The largest US category was machinery.

88. William Withington, *Crystal Palace and the World's Fair in New York City* (Lawrence, MA: n.p., 1853), NYHS Broadsides SY1853, no. 75.

89. Steen, "America's First World's Fair," 274.

90. Greeley, *Art and Industry*, 15.

91. Greeley, 30.

92. Greeley, xix.

93. Orson Desaix Munn, "Crystal Palace," *Scientific American*, October 1, 1853, 22. The sailing competition was won by a US entry named *America*.

94. Charles R. Goodrich, ed., *Science and Mechanism: Illustrated by Examples in the New York Exhibition, 1853–4* (New York: Putman, 1854), 123. Goodrich, along with Benjamin Silliman, edited *The World of Science, Art, and Industry, Illustrated from Examples in the New York Exhibition 1853–54* (New York: Putman, 1854).

95. Goodrich, *Science and Mechanism*, 129.

96. Goodrich, 177.

97. Goodrich, 157, 160, 167, 183, 150. Goodrich noted that the American Linex Thread Company, which displayed its patent linen thread at the Crystal Palace, had already taken a prize at the American Institute fair.

98. *The Diary of George Templeton Strong*, ed. Allan Nevins and Milton Halsey Thomas, vol. 2 (New York: Macmillan, 1952), 131–32, entry for October 4, 1853.

99. Orson Desaix Munn, "General Remarks," *Scientific American*, February 4, 1854, 166; Munn, "The Crystal Palace Mismanagement," *Scientific American*, April 1, 1854, 229.

100. Sarah J. Hale, "Barnum," *Godey's Lady's Book*, May 1, 1854, 469.

101. "American Museum Aerial Garden and Fair!" (New York: Applegate, 1843), American Antiquarian Society Broadside Collection.

102. "American Crystal Palace," *New Orleans Crescent*, July 22, 1854.

103. J. B. Bacon, *How to See the New York Crystal Palace: Being a Concise Guide to the Principal Objects in the Exhibition as Remodelled, 1854* (New York: G. P. Putnam, 1854), 67.

104. C. Smith, *New-York Crystal Palace: An Ode* (New York: James Egbert, printer, 1853 or 1854), NYHS Broadsides SY1854 no. 38. In addition to Smith's ode, the association offered a $200 prize for the best two odes to celebrate the reopening in 1854. Munn, "Crystal Palace Re-organized," 261. Greeley's report on the 1853 ceremonies was highly critical of the absence of working men, and even the architects were overlooked. Greeley, *Art and Industry*, 27. The reopening ceremonies and speeches were reported in the *New York Daily Tribune*, May 5, 1854.

105. Bacon, *How to See the New York Crystal Palace*, 55.

106. *New York Daily Tribune*, June 29, 1854.

107. Barnum, *Struggles and Triumphs; or Forty Years Recollections of P. T. Barnum* (Buffalo, NY: Warren, Johnson, 1873), 383; Steen, "America's First World's Fair," 285.

108. W. Byerly, "Crystal Schottisch" (New York: W. A. Pond, 1853); Emma Sampson, "Crystal Palace Cotillion" (Louisville, KY: G. W. Brainard, 1853); Frances Rziha,

"New York Crystal Palace Polka" (Boston: George P. Reed, 1853). The handkerchief is illustrated in Roberta Gorin, "In Remembrance of Things Past: Souvenirs of the Crystal Palace," in *New York Crystal Palace 1853*, digital publication, Focus Gallery exhibition at Bard Graduate Center, March 24–July 30, 2017, http://crystalpalace.visualizingnyc.org/digital-publication/in-remembrance-of-things-past-souvenirs-of-the-crystal-palace/. The original is at the New-York Historical Society.

109. Tom Standage, *The Victorian Internet* (London: Weidenfield & Nicolson, 1998); Jeffrey Kieve, *The Electric Telegraph: A Social and Economic History* (Newton Abbot: David and Charles, 1973).

110. *Harper's Weekly*, September 4, 1858.

111. Henry Kleber, "The Atlantic Telegraph Schottische" (New York: William A. Pond, 1858); Franz Kielblock, "The Ocean Cable Polka" (Boston: Oliver Ditson, 1858); A. Talexy, "Atlantic Telegraph Polka" (Boston: Oliver Ditson, 1858). "The Ocean Cable Gallop" was advertised in the *Cleveland Leader*, September 27, 1858. Philip Sharnoff, ed., *Annals of Cleveland, 1818–1935*, vol. 41 (Cleveland: Cleveland Public Library, 1937), 303.

112. *New York Daily Tribune*, September 2, 1858; *Harper's Weekly*, October 2, 1858.

113. *Harper's Weekly*, October 2, 1858.

114. *New York Daily Tribune*, September 2, 1858. The transatlantic cable remained a marvel for decades after it went into use. In 1872 *Scribner's Monthly* published "The Great Sea Serpent, A New Wonder Story," by Hans Christian Andersen. Andersen's fanciful tale depicts the surprise and consternation of sea creatures who encounter the cable lying on the ocean floor. *Scribner's Monthly*, January 1872, 325–29.

115. *New York Daily Tribune*, September 2, 1858; advertisement for Ackerman and Miller's Old Paint Shop, *New York Daily Tribune*, September 2, 1858.

116. *New York Daily Tribune*, September 2, 1858.

117. *New York Daily Tribune*, September 2, 1858.

118. *New York Daily Tribune*, September 2, 1858.

119. Atlantic Telegraph Celebration: Order of Exercises at the Crystal Palace, on the Occasion of the Celebration of the Successful Laying of the Atlantic Telegraph Cable. September 1, 1858 (New York: n.p., 1858), New-York Historical Society Broadsides SY1858 no. 72.

120. Kieve, *Electric Telegraph*, 109; Standage, *Victorian Internet*, 81.

121. Franklin Harvey Biglow to his sister Elizabeth Biglow, October 7, 1858, Museum of the City of New York, https://blog.mcny.org/2012/12/04/the-great-crystal-palace-fire-of-1858/.

122. Strong, *Diary*, 2:416, entry for October 5, 1858.

123. *Burning of the New York Crystal Palace* (New York: Currier and Ives, 1858).

124. "Crystal Palace Relics!" (New York: Wynkoop, Hallenbeck and Thomas, 1858), Museum of the City of New York collection.

125. E. H. Chaplin, *The American Idea, and What Grows Out of It: An Oration, Delivered in the New-York Crystal Palace, July 4, 1854* (Boston: A. Tompkins, 1854), 17.

126. Luther R. Marsh, *Anniversary Address before the American Institute of the City of New York at the Crystal Palace, October 25th, 1855, during the Twenty-Seventh Annual Fair, by Luther R. Marsh, Esq. of New York* (New York: Pudney and Russell, 1855), 26–27.

Conclusion

1. "An Address by the United States Centennial Commission. To the People of the United States," *Farmers' Cabinet* (Amherst, NH), November 20, 1872. Congress approved the act for the exhibition March 3, 1871. Bruno Giberti, *Designing the Centennial* (Lexington: University of Kentucky Press, 2002), 15. The authorized guidebook was the *Visitors Guide to the Centennial Exhibition and Philadelphia* (Philadelphia: J. B. Lippincott, 1876), which stated on the cover that it was "The only Guide-Book Sold on the Exhibition Grounds." The estimated total number of admissions was almost ten million. J. S. Ingram, *The Centennial Exposition Described and Illustrated* (Philadelphia: Hubbard Bros., 1876), 755.

2. "The Centennial Exhibition," *New York Times*, May 10, 1876.

3. Russell F. Weigley, ed., *Philadelphia: A 300-Year History*, 2nd ed. (New York: W. W. Norton, 1982), 429.

4. Robert C. Post, *1876: A Centennial Exhibition* (Washington, DC: Smithsonian Institution, 1976), 81; Deborah J. Warner, "The Women's Pavilion," in Post, *1876*, 165; Mary Frances Cordato, "Toward a New Century: Women and the Philadelphia Centennial Exhibition, 1876," *Pennsylvania Magazine of History and Biography* 107, no. 1 (January 1983), 125; Deborah J. Warner, "Women Inventors at the Centennial," in *Dynamos and Virgins Revisited: Women and Technological Change in History*, ed. Martha Moore Trescott (Metuchen, NJ: Scarecrow, 1979), 102–19.

5. *Visitors Guide to the Centennial Exhibition and Philadelphia*, 18.

6. Marsh, *Unravelled Dreams*, 426, 448.

7. "But there is room to hope that, before the dawn of the twentieth century, we shall be exporting instead of importing silk goods; that the moderate-priced but durable spun silks will claim their place as the most economical of dresses for our American women, while engaged in their every-day duties; and that the display of laces, ribbons, silks and velvets, greeting the eye of the visitor to the Grand Exposition which in this country shall welcome the beginning of a new century of the Christian Era." L. P. Brockett, *The Silk Industry in America: A History. Prepared for the Centennial Exhibition by L. P. Brockett, M.D.* (New York: Silk Association of America, 1876), 131–32.

8. The megatherium bones from West Virginia were identified by Casper Wister as identical to the megatherium from Paraguay identified by Georges Cuvier (and subsequently named for him). Stanley Hedeen, *Big Bone Lick: The Cradle of American Paleontology* (Lexington: University of Kentucky Press, 2008), 86.

9. Valerie Bramwell and Robert M. Peck, *All in the Bones: A Biography of Benjamin Waterhouse Hawkins* (Philadelphia: Academy of Natural Sciences of Philadelphia, 2008), 79, 80. The academy's *Hadrosaurus* was the centerpiece of its collection. The published *Guide* provided two pages of details about the dinosaur. *Guide to the Museum of the Academy of Natural Sciences of Philadelphia* (Philadelphia: published by the Academy, 1879), 20–23.

10. Orson Desaix Munn, "The Remains of a Megatherium in Ohio," *Scientific American*, January 1870, 3; Charles Betts, "The Yale College Expedition of 1870," *Harper's Monthly Magazine*, October 1871, 666.

11. Ward's 1866 catalog advertised a complete skeleton of a megatherium (unmounted) for $250. H. A. Ward, *Catalog of Casts and Fossils from the Principal*

Museums of Europe and America with Short Descriptions and Illustrations (Rochester, NY: Benton and Andrews, 1866), 14.

12. The Centennial Photographic Company's stereograph is labeled with the misspelled name *Negatherium Cavreri*. If the company did sell images of the *Hadrosaurus*, no copies exist. Frank H. Norton, ed., *Frank Leslie's Illustrated Historical Register of the Centennial Exposition 1876* (New York: Frank Leslie's Publishing House, 1876–77), 131, included an illustration of the megatherium.

13. Keith Thomson, *The Legacy of the Mastodon: The Golden Age of Fossils in America* (New Haven, CT: Yale University Press, 2008), 314.

14. James D. McCabe, *The Illustrated History of the Centennial Exhibition* (Philadelphia: National Publishing Company, 1876), 331–36.

15. McCabe, 334.

16. *Visitors Guide to the Centennial Exhibition and Philadelphia*, 37.

17. McCabe, *Illustrated History of the Centennial Exhibition*, 622.

18. Robert W. Rydell, *All the World's a Fair: Visions of Empire at American International Expositions, 1876–1916* (Chicago: University of Chicago Press, 1984), 23.

19. Rydell, 26.

20. William Dean Howells, "A Sennight of the Centennial," *Atlantic Monthly* 38, no. 225 (1876): 103.

21. Frederick Douglass's experience is detailed in Philip S. Foner, "Black Participation in the Centennial of 1876," *Phylon* 39, no. 4 (1978): 288.

22. *Visitors Guide to the Centennial Exhibition and Philadelphia*, 19; Foner, "Black Participation in the Centennial," 288.

23. David Bailey, *Eastward Ho! Leaves from the Diary of a Centennial Pilgrim* (Highland, OH: David Bailey, 1877), 48. William Dean Howells, who used his position as editor of a national publication to convey his vitriolic racism to a broad audience, told his readers that the statue was "a most offensively Frenchy negro. . . . One longs to clap him back into hopeless bondage." Howells, "Sennight of the Centennial," 98; Susanna W. Gold, *The Unfinished Exhibition: Visualizing Myth, Memory, and the Shadow of the Civil War in Centennial America* (London: Routledge, 2017), 120–21.

24. Act of July 2, 1862 (Morrill Act), Public Law 37-108; Enrolled Acts and Resolutions of Congress, 1789–1996; Record Group 11; General Records of the United States Government; National Archives, https://www.ourdocuments.gov/doc.php?flash=false&doc=33&page=transcript. Accessed April 21, 2021.

25. John Maass, "The Centennial Success Story," in *Post, 1876*, 22.

26. Maass, 23.

27. "An Address by the United States Centennial Commission. To the People of the United States," *Farmers' Cabinet* (Amherst, NH), November 20, 1872.

BIBLIOGRAPHY

Newspapers

Albany (NY) Centinel
Albany (NY) Daily Advertiser
Alexandria (VA) Gazette Commercial and Political
American Beacon and Commercial Diary (Norfolk, VA)
American Beacon and Norfolk-Portsmouth (VA) Daily Advertiser
American Citizen and General Advertiser (New York)
American Daily Advertiser (Philadelphia)
American Herald (Boston)
American Journal (Ithaca, NY)
American Mercury (Hartford, CT)
Aurora General Advertiser (Philadelphia)
Baltimore Gazette and Daily Advertiser
Baltimore Patriot and Evening Advertiser
Baltimore Patriot and Mercantile Advertiser
Barre (MA) Gazette
Barre (MA) Patriot
Binghamton (NY) Phoenix
Boston Commercial Gazette
Boston Daily Advertiser
Boston Daily Atlas
Boston Evening-Post
Boston Gazette, or Weekly Journal
Boston Intelligencer and Evening Gazette
Boston Intelligencer and Morning and Evening Advertiser
Bruce Herald (Milton, New Zealand)
Cherry Valley (NY) Gazette
City Gazette and Daily Advertiser (Charleston, SC)
Cleveland Leader
Cleveland Plain Dealer
Columbian (New York)
Columbian Centinel (Boston)

Columbian Phenix or Providence (RI) Patriot
Columbian Register (New Haven, CT)
Connecticut Courant (Hartford)
Connecticut Herald (New Haven)
Connecticut Journal (Hartford)
Daily Advertiser (Newark, NJ)
Daily Evening Transcript (Boston)
Daily Madisonian (Washington, DC)
Daily National Intelligencer (Washington, DC)
Democratic Press (Philadelphia)
Democratic Republican (Philadelphia)
Diary or Loudon's Register (New York)
Downfall of Babylon, or, The Triumph of Truth over Popery (New York)
Dunlap's American Daily Advertiser (Philadelphia)
Eastern Argus (Portland, ME)
Easton (MD) Gazette and Eastern Shore Intelligencer
Enquirer (Richmond, VA)
Essex Register (Salem, MA)
Falmouth (ME) Gazette and Weekly Advertiser
Farmer's Cabinet (Amherst, NH)
Federal Gazette and Philadelphia Daily Advertiser
Frank Leslie's Illustrated Newspaper (New York)
Frederick Douglass' Paper (Rochester, NY)
Freeman's Journal: or, The North-American Intelligencer (Philadelphia)
Gazette of the United States (Philadelphia)
Georgia Gazette (Savannah)
Hampshire Gazette (Northampton, MA)
Harper's Weekly (New York)
Hudson River Chronicle (Ossining, NY)
Independent Chronicle and Universal Advertiser (Boston)
Independent Gazetteer (Worcester, MA)
Independent Gazetteer (Philadelphia)
Independent Journal (New York)
Independent Ledger and the American Advertiser (Boston)
Jamestown (NY) Journal
Maryland Journal and Baltimore Advertiser
Massachusetts Centinel and the Republican Journal (Boston)
Massachusetts Gazette (Boston)
Massachusetts Spy (Boston)
Mercantile Advertiser (New York)

Milwaukee Daily Sentinel
National Advocate (New York)
National Anti-Slavery Standard (New York)
National Standard (Middlebury, VT)
Newark (NJ) Observer
Newburyport (MA) Herald and Country Gazette
New-England Courant (Boston)
New-Hampshire Gazette and General Advertiser (Portsmouth)
New Hampshire Patriot and State Gazette (Concord)
New-Hampshire Sentinel (Keene)
New Haven (CT) Daily Herald
New-Haven Gazette and Connecticut Magazine
New-Jersey Gazette (Trenton)
New Orleans Crescent
Newport (RI) Mercury
New York Columbian
New York Commercial Advertiser
New York Daily Times
New York Daily Tribune
New York Evening Post
New-York Gazette and General Advertiser
New-York Gazette, or, the Weekly Post-Boy
New York Herald
New York Journal
New-York Mercury
New York Morning Herald
New York Packet and American Advertiser
New York Times
New York Sun
Niles Weekly Register (Baltimore)
Northern Whig (Hudson, NY)
Norwich (CT) Packet and Country Journal
Pennsylvania Gazette (Philadelphia)
Pennsylvania Journal (Philadelphia)
Pennsylvania Packet, and Daily Advertiser (Philadelphia)
Pennsylvania Packet, and General Advertiser (Philadelphia)
Philadelphia Gazette and Universal Daily Advertiser
Philadelphia Repository, and Weekly Register
Pittsfield (MA) Sun
Ploughman; or, Republican Federalist (Bennington, VT)

Plough Boy (Albany, NY)

Political Intelligencer and New-Jersey Advertiser (New Brunswick)

Port Folio (Philadelphia)

Poulson's American Daily Advertiser (Philadelphia)

Public Ledger (Philadelphia)

Relf's Philadelphia Gazette, and Daily Advertiser

Repertory (Boston)

Republican or, Anti-Democrat (Baltimore)

Republican Watch-Tower (New York)

Rhode Island American, Statesman and Providence Gazette

Richmond (VA) Enquirer

Royal Gazette (New York)

Salem (MA) Gazette

Savannah Republican

South Carolina Gazette and General Advertiser (Charleston)

Southern Patriot (Charleston, SC)

Spectator (New York)

Springfield (MA) Republican

State Gazette (Austin, TX)

State Gazette of South Carolina (Charleston)

Telescope (Columbia, SC)

Tickler (Philadelphia)

Times and Weekly Adviser (Hartford, CT)

Union Herald (Cazenovia, NY)

Virginia Journal and Alexandria Advertiser

Washington (DC) Sentinel

Western Star (Stockbridge, MA)

Wilmington (NC) Journal

Primary Sources

Abbott, Jacob. *Rollo's Travels*. Philadelphia: Hogan and Thompson; Boston: Gould, Kendall and Lincoln, 1839.

"Abstract of the Report of the Committee on Premiums and Exhibition on the Subject of the Third Annual Exhibition, and of the Premiums Awarded." *Franklin Journal and American Mechanics Magazine* 2, no. 5 (November 1826): 264–68.

"Account and Conditions of a Premium Offered by the American Philosophical Society." *Columbian Magazine* 1, no. 4 (December 1786): 179–80.

American Academy of Arts and Sciences. "Charter of Incorporation, May 4, 1780." In *Memoirs of the American Academy of Arts and Sciences* 1 (1783): iv–vii.

American Academy of Arts and Sciences. *Memoirs of the American Academy of Arts and Sciences* 1 (1783). Boston: Adams and Nourse, 1785.

American Academy of Arts and Sciences. Preface to *Memoirs of the American Academy of Arts and Sciences* 1 (1783): iii–xi.

American Philosophical Society. *Transactions of the American Philosophical Society, Held at Philadelphia, for Promoting Useful Knowledge.* Vol. 1. Philadelphia: William and Thomas Bradford, 1771.

American Philosophical Society. *Transactions of the American Philosophical Society, Held at Philadelphia, for Promoting Useful Knowledge.* Vol. 1. 2nd ed. Philadelphia: William and Thomas Bradford, 1789.

An Account of Count D'Artois and his Friend's Passage to the Moon in a Flying Machine, called, An Air Balloon. Litchfield, CT: Collier and Capp, 1785.

Address of the Committee on Premiums and Exhibitions of the Franklin Institute. Philadelphia: J. Harding, 1832.

Andersen, Hans Christian. "The Great Sea Serpent, a New Wonder Story." *Scribner's Monthly,* January 1872, 325–29.

Antill, Edward. "An Essay on the Cultivation of the Vine, and the making and preserving of Wine, suited to the different Climates in North-America." *Transactions of the American Philosophical Society* 1 (2nd ed., 1789): 180–262.

Appleton, George S. *A Handbook for the Stranger in Philadelphia.* Philadelphia: George S. Appleton, 1849.

"Article V. Character of Eustache." *American Phrenological Journal and Miscellany* 2, no. 4 (January 1, 1840): 177–82.

Atlantic Telegraph Celebration: Order of Exercises at the Crystal Palace, on the Occasion of the Celebration of the Successful Laying of the Atlantic Telegraph Cable. September 1, 1858. New York: n.p. New York Historical Society Broadsides SY1858 no. 72.

Aunt Mary's Stories for Children. New York: Leavitt and Allen, 1852.

Bacon, J. B. *How to See the New York Crystal Palace: Being a Concise Guide to the Principal Objects in the Exhibition as Remodelled, 1854.* New York, G. P. Putnam, 1854.

Bailey, David. *Eastward Ho! Leaves from the Diary of a Centennial Pilgrim.* Highland, OH: David Bailey, 1877.

Baldwin, Christopher Columbus. *A Place in My Chronicle: A New Edition of the Diary of Christopher Columbus Baldwin, 1829–1835.* Edited by Jack Larkin and Caroline Sloat. Worcester, MA: American Antiquarian Society, 2010.

Baldwin, Loammi, Jr. *Report on Introducing Pure Water into the City of Boston.* 2nd ed. Boston: Hilliard, Gray, 1835.

"The Balloon." In *The Evergreen: or, Stories for Childhood and Youth,* edited by Walter West, 15–20. Boston: Munroe and Francis, 1847.

Banneker, Benjamin. *Benjamin Banneker's Pennsylvania, Delaware, Maryland and Virginia Almanac and Ephemeris for the Year of Our Lord, 1792.* Baltimore: William Goddard and James Angell, 1791.

Barber, John Warner. *A History of the* Amistad *Captives.* New Haven, CT: E. L. and J. W. Barber, 1840.

Barnum, P. T. *Struggles and Triumphs; or, Forty Years Recollections of P. T. Barnum.* Buffalo, NY: Warren, Johnson, 1873.

Bartram, Moses. "Observations on the Native Silk-Worms of North America." In *Transactions of the American Philosophical Society* 1 (1771): 294–301.

Bentley, William. *The Diary of William Bentley, Pastor of the East Church, Salem, Massachusetts.* Vol. 4. Salem, MA: Essex Institute, 1914.

Betts, Charles. "The Yale College Expedition of 1870," *Harper's Monthly Magazine,* October 1871, 666.

Blake, George. *Blake's Collection of Duetts for Two Flutes, Clarinets, or Violins.* Philadelphia: George E. Blake, 1807.

Bowen, Eli. *The Pictorial Sketch-Book of Pennsylvania.* Philadelphia: William Bromwell, 1852.

Breck, Samuel. "The Diary of Samuel Breck, 1814–1822." Edited by Nicholas B. Wainwright. *Pennsylvania Magazine of History and Biography* 102, no. 4 (October 1978): 469–508.

Brewer, George. *A Description of the Mammoth Cave of Kentucky and Niagara River and Falls, and the Falls in Summer and Winter; the Prairies, or Life in the West; the Fairmount Water Works and Scenes on the Schuylkill, &c. &c. To Illustrate Brewer's Panorama.* Boston: J. M. Hewes, 1859.

Brockett, L. P. *The Silk Industry in America: A History. Prepared for the Centennial Exhibition by L. P. Brockett, M.D.* New York: Silk Association of America, 1876.

Brown, William H. *The History of the First Locomotives in America from the Original Documents, and the Testimony of Living Witnesses.* New York: D. Appleton, 1874.

Burges, Tristram. *Address of the Hon. Tristram Burges [of Rhode Island], Delivered at the Third Annual Fair of the American Institute of the City of New York. Held at Masonic Hall, October, 1830.* New York: John M. Danforth, 1830.

Burr, Esther Edwards. *The Journal of Esther Edwards Burr, 1754–1757.* Edited by Carol F. Karlsen and Laurie Crumpacker. New Haven, CT: Yale University Press, 1984.

Caldwell, Charles. *Elements of Phrenology.* Lexington, KY: A. G. Meriwether, printer, 1827.

Capen, Nahum. *Reminiscences of Dr. Spurzheim and George Combe.* New York: Fowler and Wells, 1881.

Celebration of the Introduction of the Water of Lake Cochituate into the City of Boston, October 25, 1848. Boston: J. H. Eastburn, 1848.

"Centre Square Philad'a." In *The Casket; or, Flowers of Literature, Wit and Sentiment.* Philadelphia:, n.p., October 1831.

Chaplin, E. H. *The American Idea, and What Grows Out of It: An Oration, Delivered in the New-York Crystal Palace, July 4, 1854.* Boston: A. Tompkins, 1854.

Child, Lydia Maria. *Flowers for Children.* Boston: Crosby and Nichols, 1854.

Child, Lydia Maria. *Letters from New York.* New York: Charles S. Francis, 1843.

Clemens, Samuel L., and Charles Neider. *The Autobiography of Mark Twain.* New York: Harper, 1959.

Colden, Cadwallader D. *The Life of Robert Fulton.* New York: Kirk and Mercein, 1817.

Colfax, Richard. *Evidence against the Views of the Abolitionists, Consisting of Physical and Moral Proofs of the Natural Inferiority of the Negroes.* New York: James T. M. Bleakley, 1833.

Collyer, Robert H. *Lights and Shadows of American Life.* Boston: Redding, 1843.

Colman, Samuel. *The Child's Gem.* New York: S. Colman, 1839.

Combe, Andrew. "Remarks on Tiedemann's Comparison of the Negro Brain and Intellect with Those of the European." *Eclectic Journal of Medicine* (Philadelphia) 2, no. 9 (July 1838): 325–28.

Combe, George. "Chapter II: The Characteristics of Races, Masses, and Nations, in Part Hereditary, Section I. The Colored Race." *American Phrenological Journal and Miscellany*, September 1, 1843: 417–32.

Combe, George. *The Constitution of Man Considered in Relation to External Objects.* Edinburgh: John Anderson, 1828.

Combe, George. *Elements of Phrenology.* 7th ed. Edinburgh: Maclachlan and Stewart, 1850.

Combe, George. *Notes on the United States of North America, during a Phrenological Visit in 1838–9–40.* 3 vols. Edinburgh: Maclachlan, Stewart, 1841.

Combe, George. *A System of Phrenology.* 3rd ed. London: Longman, 1830.

Cope, Thomas. *Philadelphia Merchant: The Diary of Thomas P. Cope, 1800–1851.* Edited by Eliza Cope Harrison. South Bend, IN: Gateway, 1978.

Coxe, Tench. *An Address to the Assembly of the Friends of American Manufactures.* Philadelphia: Robert Aiken and Sons, 1787.

Coxe, Tench. *Observations on the Agriculture, Manufactures and Commerce of the United States.* New York: Francis Childs and John Swaine, 1789.

Denon, Baron Dominique Vivant. *Voyage dans la Basse et la Haute Égypte pendant les campagnes du général Bonaparte.* Paris: P. Didot l'aîné, 1802.

"A Description of a New Invention of a Spinning Machine." *Pennsylvania Magazine, or, American Monthly Museum* 1 (April 1, 1775): 158–59.

Dickens, Charles. *American Notes for General Circulation.* London: Chapman and Hall, 1842. Reprint, New York: Penguin, 1985.

Draper, Richard. *Blazing-stars messengers of God's wrath: in a few serious and solemn meditations upon the wonderful comet* [. . .]. Boston: Printed and sold by R. Draper in Newbury-Street; and by Fowle and Draper in Marlborough-Street, 1759.

Drinker, Elizabeth. *The Diary of Elizabeth Drinker.* 3 vols. Edited by Elaine Forman Crane. Boston: Northeastern University Press, 1991.

Dwight, Timothy. *Travels in New-England and New York.* Vol. 1. New Haven, CT: Timothy Dwight, 1821.

Elisha, Patrick N. I. [Oliver Evans]. *Patent Right Oppression Exposed, or Knavery Detected.* Philadelphia: R. Folwell, 1813.

Ellis, Edward S. *The Steam Man of the Prairies. Beadle's American Novel*, no. 45, August 1868.

Emerson, Ralph Waldo. "Historic Notes on Life and Letters in New England." In *The Complete Works of Ralph Waldo Emerson*, vol. 10, edited by Edward Waldo Emerson, 337–39. Boston: Houghton, Mifflin, 1904.

Evans, Oliver. *The Abortion of the Young Steam Engineer's Guide.* Philadelphia: Fry and Kammerer, 1805.

Evans, Oliver. *Oliver Evans to His Counsel: Who Are Engaged in Defence of His Patent Rights, for the Improvements He Has Invented.* Philadelphia: Oliver Evans, 1817.

Everett, Edward. *Address Delivered before the American Institute of the City of New York at The Fourth Annual Fair, October 14, 1831.* New York: Van Norden and Mason, 1831.

Everett, Edward. *A Memoir of Mr. John Lowell, Junior.* Boston: Little, 1840.

Ewbank, Thomas. *A Descriptive and Historical Account of Hydraulic and Other Machines for Raising Water*. 4th ed. New York: Greeley & McElrath, 1850.

"Extract from First Annual Report of the Pennsylvania Society for the Promotion of Internal Improvements in the Commonwealth." *Franklin Journal and American Mechanics Magazine* 1, no. 1 (January 1826): 11–15.

First Annual Report of the Proceedings of the Franklin Institute of the State of Pennsylvania, for the Promotion of the Mechanic Arts. Philadelphia: J. Harding, printers, 1825.

Fowler, Lorenzo. "Phrenological Developments of Joseph Cinquez, Alias Ginqua." *American Phrenological Journal and Miscellany* 2, no. 3 (December 1839): 136–38.

Fowler, Lorenzo. "A Shelf in Our Cabinet—No. 2." *American Phrenological Journal* 35, no. 3 (March 1862): 52.

Fowler, Lorenzo. "Slavery Preferred to Freedom." *American Phrenological Journal and Miscellany*, November 1, 1850: 359.

Fowler, Orson. "Sarah Kinson, or Margru." *American Phrenological Journal* 12, no. 7 (July 1, 1850): 230.

Fowler, Orson. *Human Science or Phrenology*. Philadelphia: National, 1873.

Fowler, Orson, and Lorenzo Fowler. "A List of Specimens Designed for Phrenological Societies." *American Phrenological Journal and Miscellany*, April 1, 1848: 129–30.

Franklin, Benjamin. *The Papers of Benjamin Franklin*. New Haven, CT: Yale University Press, 1959.

Franklin, Benjamin. *Poor Richard Improved; Being an Almanack* [. . .] *for the Year of Our Lord 1753* [. . .]. Philadelphia: Franklin and Hall, 1752.

Fulton, Robert. *Torpedo War and Submarine Explosions*. New York: William Elliot, 1810.

Gale, Benjamin. "Observations on the Culture of Smyrna Wheat," *Memoirs of the American Academy of Arts and Sciences* 1 (1783): 381–82.

Garfield, James A. *The Diary of James A. Garfield*. Edited by Harry James Brown and Frederick D. Williams. Vol. 1. East Lansing: Michigan State University Press, 1967.

Goodrich, Charles R., ed. *Science and Mechanism: Illustrated by Examples in the New York Exhibition, 1853–4*. New York: Putman, 1854.

Goodrich, Samuel G. "The Crystal Palace in New York." *Merry's Museum and Parley's Magazine*, January 1853, 63–64. New York: S. T. Allen.

Goodrich, Samuel G. *Robert Merry's Museum*. Vol. 6. New York: G. W. and S. O. Post, 1847.

A Grammar of Chemistry Wherein the Principles of the Science Are Familiarized by a Variety of Easy and Entertaining Experiments with Questions for Exercise, and a Glossary of Terms in Common Use by D. Blair Corrected and Revised by Benjamin Tucker. Philadelphia: David Hogan, 1810.

The Great Steam Duck; or, A Concise Description of a Most Useful and Extraordinary Invention for Aerial Navigation, Authored by an Anonymous "Member of the L.L.B.B." Louisville, KY: Henkle, Logan, 1841.

Greeley, Horace. *Art and Industry as Represented at the New York Crystal Palace*. New York: Putnam, 1853.

Greeley, Horace. *Glances at Europe in a Series of Letters*. New York: Dewitt and Davenport, 1852.

Greenleaf, Joseph. "An Account of an Experiment for raising Indian Corn, in poor Land." *Memoirs of the American Academy of Arts and Sciences* 1 (1783): 383–85.

Guide to the Museum of the Academy of Natural Sciences of Philadelphia. Philadelphia: published by the Academy, 1879.

Hale, Sarah J. "Barnum." *Godey's Lady's Book*, May 1, 1854, 469.

Halleck, Fitz-Greene. *Fanny*. New York: C. Wiley, 1819.

A Handbook for the Stranger in Philadelphia. Philadelphia: George S. Appleton, 1849.

Hawthorne, Nathaniel. *Nathaniel Hawthorne: The American Notebooks*. Edited by Claude M. Simpson. Columbus: Ohio State University Press, 1972.

Hawthorne, Nathaniel. "Phrenology." *American Magazine of Useful and Entertaining Knowledge*, April 1, 1836, 337–38.

Hone, Philip. *The Diary of Philip Hone, 1828–1851*. Edited by Bayard Tuckerman. Vol. 2. New York: Dodd, Mead, 1889.

Howells, William Dean. "A Sennight of the Centennial." *Atlantic Monthly* 38, no. 225 (1876): 92–107.

Ingram, J. S. *The Centennial Exposition Described and Illustrated*. Philadelphia: Hubbard Bros., 1876.

Jackson, Andrew. Fifth Annual Message to Congress, December 3, 1833. American Presidency Project. https://www.presidency.ucsb.edu/documents/fifth-annual-message-2.

James (a young correspondent). "Mr. Durant." *Parley's Magazine*, September 14, 1833, 18.

"January First, A.D. 3000." *Harper's New Monthly Magazine*, January 1856, 145–57.

Jefferson, Thomas. *Notes on the State of Virginia*. 1787. University of Virginia Library, Electronic Text Center. Accessed March 12, 2019. http://etext.lib.virginia.edu/toc/modeng/public/JefVirg.html.

Jefferson, Thomas. *The Papers of Thomas Jefferson Digital Edition*. Edited by Barbara B. Oberg and J. Jefferson Looney. Charlottesville: University of Virginia Press, Rotunda, 2008. Accessed September 13, 2009. http://rotunda.upress.virginia.edu:8080/founders/default.xqy?keys=TSJN-print-01-22-02-0091.

Journal of the Twenty Third House of Representatives of the Commonwealth of Pennsylvania. Harrisburg, PA: J. Peacock, printer, 1812.

King, Charles. *A Memoir of the Construction, Cost, and Capacity of the Croton Aqueduct [. . .] Together with an Account of the Civic Celebration of the Fourteenth October, 1842, on Occasion of the Completion of the Great Work*. New York: printed by C. King, 1843.

Kotzebue, August von. Lieut. Col. Capadose, translator. *The Organs of the Brain, a Comedy in Three Acts*. London: Edward Bull, 1838.

Lackington, James. *Memoirs of the Forty-Five Years of the Life of James Lackington*. London: printed for the author, 1827.

Landell, E. *The Boy's Own Toy-Maker: A Practical Illustrated Guide to the Useful Employment of Leisure Hours*. Boston: Shepard, Clark and Brown, 1859.

Lardner, Dionysius. *Popular Lectures on Science and Art: Delivered in the Principal Cities and Towns of the United States*. New York: Greeley and McElrath, 1845.

Lardner, Dionysius. *Popular Lectures on the Steam Engine, in Which Its Construction and Operation Are Familiarly Explained: With an Historical Sketch of Its Invention and Progressive Improvement with Additions by James Renwick.* New York: printed for E. Bliss, 1828.

Latrobe, Benjamin Henry. *Anniversary Oration, Pronounced before the Society of Artists of the United States.* Philadelphia: Bradford and Inskeep, 1811.

Latrobe, Benjamin Henry. *View of the Practicality and Means of Supplying the City of Philadelphia with Wholesome Water in a Letter to John Miller, Esquire, December 29th, 1798.* Philadelphia: Zachariah Poulson Jr., 1799.

Leeds, Daniel. *The American Almanack for the Year of Christian Account, 1712.* New York: William and Andrew Bradford, 1712.

Logan, Martha, and Mary Barbot Prior. "Letters of Martha Logan to John Bartram, 1760–1763." *South Carolina Historical Magazine* 59, no. 1 (January 1958): 38–46.

Mabbott, Thomas Ollive. *See* Poe, Edgar Allan.

Manigault, Ann. "Extracts from the Journal of Mrs. Ann Manigault 1754–1781," *South Carolina Historical and Genealogical Magazine* 20 (1919): 57–63, 128–41, 204–12, 254–59.

Marsh, Luther R. *Anniversary Address before the American Institute of the City of New York at the Crystal Palace, October 25th, 1855, during the Twenty-Seventh Annual Fair, by Luther R. Marsh, Esq. of New York.* New York: Pudney and Russell, 1855.

Martin, Benjamin. *Supplement Containing Remarks on a Rhapsody of Adventures of a Modern Knight-errant.* Bath, UK: printed for the author, 1746.

Martineau, Harriet. *Retrospect of Western Travel.* Vol. 2. London: Saunders and Otley, 1838.

Mason, Monck. *Account of the late Aeronautical Expedition from London to Weilburg.* New York: T. Foster, 1837.

Mather, Cotton. "Advice from Taberah: A Sermon Preached after the Terrible Fire." Boston: B. Green, 1711.

Mather, Increase. *Kometographia, or, A discourse concerning comets; wherein the nature of blazing stars is enquired into: with an historical account of all the comets which have appeared from the beginning of the world unto this present year, M.DC.LXXXIII [. . .] : as also two sermons occasioned by the late blazing stars.* Boston: Samuel Green, 1683.

Mayhew, Jonathan. "God's Hand and Providence to be religiously acknowledged in public Calamities: A Sermon Occasioned by the Great Fire in Boston, New England." Boston: Richard Draper, 1760.

McCabe, James D. *The Illustrated History of the Centennial Exhibition.* Philadelphia: National Publishing Company, 1876.

Melville, Herman. *The Confidence Man.* New York: Dix, Edwards, 1857.

Melville, Herman. *Moby-Dick; or, The Whale.* New York: Harper and Brothers, 1851.

Meredith, David Reese. *Humbugs of New-York; Being a Remonstrance against Popular Delusion; Whether Science, Philosophy, or Religion.* New York: Taylor, 1838.

M'Henry, James. *An Ode: Written by Request, on the Opening of the Exhibition at the Franklin Institute of Philadelphia, October, 1828.* Philadelphia: n.p., 1828.

Mickle, Samuel. *The Diaries of Samuel Mickle: Woodbury, Gloucester County, New Jersey, 1792–1829.* Transcribed by Ruthe Baker. Woodbury, NJ: Gloucester County Historical Society, 1991.

Miller, Geoffrey. "The Alida Waltz: Early 19th Century Music and the Hudson River." Hudson River Maritime Museum website. Accessed April 16, 2020. http://www.hrmm.org/history-blog/the-alida-waltz-early-19th-century-musicand-the-hudson-river.

Morton, Henry B. "The Redheffer Perpetual Motion Machine." *Journal of the Franklin Institute* (Philadelphia) 139 (1895): 246–51.

Morton, Samuel George. *Crania Americana; or, A Comparative View of the Skulls of Various Aboriginal Nations of North and South to Which Is Prefixed an Essay on the Varieties of the Human Species.* Philadelphia: J. Dobson, 1839.

Munn, Orson Desaix. "Crystal Palace." *Scientific American*, October 1, 1853, 22.

Munn, Orson Desaix. "The Crystal Palace Mismanagement." *Scientific American*, April 1, 1854, 229.

Munn, Orson Desaix. "The Crystal Palace Re-organized." *Scientific American*, April 19, 1854, 261.

Munn, Orson Desaix. "General Remarks." *Scientific American*, February 4, 1854, 166.

Munn, Orson Desaix. "The Remains of a Megatherium in Ohio." *Scientific American*, January 1870, 3.

Nason, Elias. *Sir Charles Henry Frankland, Baronet; or, Boston in the Colonial Times.* Albany, NY: J. Munsell, 1865.

National Magazine; Devoted to Literature, Art, and Religion (New York), January 1853.

Neal, John. "The Young Phrenologist." *New England Galaxy and United States Literary Advertiser*, October 3, 1835.

"New-York Crystal Palace for the Exhibition of the Industry of All Nations." *Plough, the Loom, and the Anvil*, November 1852, 315.

Nollet, Jean-Antoine. *Leçons de physique expérimentale.* Vol. 5. Paris: H. H. Guerin & L. F. Delatour, 1777.

Norton, Frank H., ed. *Frank Leslie's Illustrated Historical Register of the Centennial Exposition 1876.* New York: Frank Leslie's Publishing House, 1876–77.

"Observations on the Rise and Progress of the Franklin Institute." *Franklin Journal and American Mechanics Magazine* 1, no. 2 (February 1826): 66–70.

O'Connor, Thomas H. *The Athens of America: Boston, 1825–1845.* Amherst, MA: University of Massachusetts Press, 2006.

"An Ode, Sung by Capt. Joseph Cowdin, at the Supper of the Tenth Annual Fair of the American Institute. Composed by Rufus Dawes, for the Occasion." *Journal of the American Institute of the City of New York* 3, no. 1 (1838): 24.

Official Catalog of the Great Exhibition of the Works of Industry of All Nations, 1851. London: Spicer Brothers, 1851.

Olmstead, E. B. "George, A Slave, Murderer of Mrs. Foster, with a View of His Skull." *American Phrenological Journal and Miscellany*, November 1, 1849: 340–41.

"On the American Scheme of Establishing Colonies of Free Negro Emigrants on the Coast of Africa, as Exemplified in Liberia." *Annals of Phrenology*, October 1, 1833: 113–33.

Paine, Thomas. "Philosophical Queries." *Pennsylvania Magazine, or, American Monthly Museum* 1 (August 1, 1775): 353.

Parkes, Samuel. *A Chemical Catechism for the Use of Young People.* London: printed for the author, 1806.

Parley, Peter, ed. *Balloon Travels of Robert Merry and His Young Friends over Various Countries in Europe.* New York: J. C. Derby, 1856.

Parley, Peter. *Parley's Magazine*, December 20, 1834, 5–6.

Parley, Peter. *Parley's Magazine for Children and Youth*, August 31, 1833, 85–89.

Paulding, James Kirke. *The Merry Tales of the Three Wise Men of Gotham.* New York: G. and C. Carvill, 1826.

Peale, Charles Willson. *Introduction to a Course of Lectures on Natural History.* Philadelphia: Francis and Robert Bailey, 1800.

Phelps' New York City Guide; Being a Pocket Directory for Strangers and Citizens to the Prominent Objects of Interest in the Great Commercial Metropolis and Conductor to Its Environs. New York: Ensign, Bridgman, and Fanning, 1854.

A Phrenological Chart: Presenting a Synopsis of the Doctrine of Phrenology. Also, an Analysis of the Fundamental Powers of the Human Mind [. . .] Together with the Phrenological Character of [blank] / *Examined by Rev. Josiah M. Graves.* Hartford, CT: Hurlbut and Williams, printers, 1839.

"Phrenological Developments and Character of the Celebrated Indian Chief and Warrior, Black Hawk; With Cuts." *American Phrenological Journal and Miscellany* 1, no. 2 (November 1, 1838): 51–61.

"Phrenology." *American Magazine of Useful and Entertaining Knowledge*, April 1, 1836, 337–38.

"Phrenology Made Easy." *Knickerbocker Magazine*, June 2, 1838.

Poe, Edgar Allan. *The Collected Works of Edgar Allan Poe.* Vol. 3. Cambridge, MA: Harvard University Press, 1978. https://www.eapoe.org/works/mabbott/tom3t015.htm.

Poe, Edgar Allan. "Critical Notices." *Southern Literary Messenger* 2, no. 3 (March 1836): 286–87.

Poe, Edgar Allan. *Essays and Reviews.* New York: Library of America, 1984.

Poe, Edgar Allan. "The Murders in the Rue Morgue." *Graham's Magazine*, April 1841, 166–79.

Poe, Edgar Allan. "The Unparalleled Adventure of One Hans Pfaall." *Southern Literary Messenger* 1, no. 10 (June 1835): 565–80.

Porter, Rufus. *Aerial Navigation: The Practicality of Traveling Pleasantly and Safely from New-York to California in Three Days.* New York: H. Smith, 1849.

Porter, Rufus. "To the American Public." *Scientific American*, August 28, 1845, 2.

Pownall, Thomas. "A Brief." Boston: Richard Draper, 1760.

Premiums Awarded by the Managers of the Twenty-Sixth Annual Fair of the American Institute, 351 Broadway, October, 1853. New York: W. H. Tinson, 1854.

Priestley, Joseph. *A Familiar Introduction to the Study of Electricity.* London: J. Johnson, 1768.

Priestley, Joseph. *History and Present State of Electricity.* London: J. Dodsley, 1767.

"Proposals of the Franklin Institute, for the Exhibition of October 1826, Addressed to the Manufacturers of the United States." *Franklin Journal and American Mechanics Magazine* 1, no 1 (January 1826): 5–6.

"Quarterly Meeting of the Franklin Institute." *Franklin Journal and American Mechanics Magazine* 2, no. 5 (November 1826): 268–69.

Remarks on the Ellipsoidal Balloon, Propelled by the Archimedean Screw, Described as the New Aerial Machine. London: Howlett and Son, 1843.

Report of the Third Annual Fair of the American Institute of the City of New York Held at Masonic Hall, October, 1830 [. . .]. New York: J. Seymour, 1830.

Richards, William. *A Day in the Crystal Palace and How to Gain the Most of It.* New York: G. P. Putnam, 1853.

"Report on Instruction from Board of Managers Eighth Quarterly Report." *Franklin Journal and American Mechanics Magazine* 1, no. 1 (January 1826): 3–4.

Roberts, Solomon R. *The Promotion of the Mechanic Arts in America: An Address Delivered at the Close of the Sixteenth Exhibition of American Manufactures, Held in Philadelphia, by the Franklin Institute of the State of Pennsylvania for the Promotion of the Mechanic Arts, October 30, 1846.* Philadelphia: printed by John C. Clark, 1846.

Rodgers, Charles T. *American Superiority at the World's Fair.* Philadelphia: J. J. Hawkins, 1852.

Rush, Benjamin. *The Autobiography of Benjamin Rush: His "Travels through Life" together with His Commonplace Book for 1789–1813.* Edited by George W. Corner. Princeton, NJ: Princeton University Press, 1948.

Rush, Benjamin. *Letters of Benjamin Rush.* Vol. 2, *1793–1813.* Edited by L. H. Butterfield. Princeton, NJ: Princeton University Press, 1951.

Rush, Benjamin. *Syllabus of lectures containing the application of the principles of natural philosophy, and chemistry, to domestic and culinary purposes: Composed for the use of the Young Ladies' Academy, in Philadelphia.* Philadelphia: printed for Andrew Brown, 1787.

Rush, Benjamin. "Thoughts upon Female Education, accommodated to the Present State of Society, manners, and Government, in the United States of America [. . .]." Philadelphia, 1791.

Sellers, George Escol. *Early Engineering Reminiscences (1815–40) of George Escol Sellers.* Edited by Eugene S. Ferguson. *Smithsonian Institution Bulletin* 238. Washington, DC: Smithsonian Institution, 1965.

Sheldon, John P. "A Description of Philadelphia in 1825." *Pennsylvania Magazine of History and Biography* 60, no. 1 (January 1936): 74–76.

Shepherd, Job. *Poor Job, 1753.* Newport, RI: James Franklin, 1752.

Sherman, Roger. *An Astronomical Diary.* New York: Henry De Foreest, 1752.

Silliman, Benjamin, and Charles R. Goodrich, eds. *The World of Science, Art, and Industry, Illustrated from Examples in the New York Exhibition 1853–54.* New York: Putman, 1854.

Sizer, Nelson. *What to Do, and Why; and How to Educate Each Man for His Proper Work.* New York: Mason, Baker and Pratt, 1872.

Smith, Abigail A., and C. A. S. De Windt. *Journal and Correspondence of Miss Adams, Daughter of John Adams, Second President of the United States: Written in France and England in 1785.* Vol. 1. New York: Wiley and Putnam, 1841.

Smith, Charles. *The New-York Crystal Palace: An Ode.* New York: James Egbert, printer, [1853 or 1854].

Smith, Samuel Stanhope. *An Essay on the Causes of the Variety of Complexion and Figure in the Human Species.* New Brunswick, NJ: J. Simpson, L. Deare, printer, 1810.

Smith, William. "An Account of the Transit of Venus over the Sun's Disk, as observed at Norriton, in the County of Philadelphia, and Province of Pennsylvania, June 3rd, 1769." *Transactions of the American Philosophical Society* 1 (1771): 24.

Spurzheim, J. G. *Phrenology in Connexion with the Study of Physiognomy.* Boston: Marsh, Capen and Lyon, 1833.

Stevens, Abel. "The New York Crystal Palace." *National Magazine: Devoted to Literature, Art, and Religion* (New York), January 1853, 80–81.

Stowe, Harriet Beecher. "The Canal Boat." *Godey's Lady's Book*, 1841.

Stowe, Harriet Beecher. *Uncle Tom's Cabin; or, Life among the Lowly* (Boston: John P. Jewett, 1852).

The Stranger's Guide to the Public Buildings, Places of Amusement, Streets, Lanes, Alleys, Roads, Avenues, Courts, Wharves, Principal Hotels, Steam-Boat Landings, Stage Offices, Etc. Etc. of the City of Philadelphia and Adjoining Districts. Philadelphia: H. S. Tannes, 1828.

Strong, George Templeton. *The Diary of George Templeton Strong.* Edited by Allan Nevins and Milton Halsey Thomas. 3 vols. New York: Macmillan, 1952.

Swift, Mary A. *First Lessons on Natural Philosophy for Children, Part Second.* Hartford, CT: Belknap and Hamersley, 1836.

Thinker, Theodore. *The Balloon and Other Stories.* New York: Clark, Austin, 1851.

Thistlewood, Thomas. Diary. Lincolnshire County Archives, Lincoln, UK.

Transactions of the American Philosophical Society. See American Philosophical Society.

"The Travelling Balloon." *Scientific American*, March 3, 1849, 189.

Trollope, Frances. *Domestic Manners of the Americans.* Vol. 2. London: Whittaker, Treacher, 1832.

Twain, Mark. *Tom Sawyer.* Hartford, CT: American Publishing, 1876.

United States Patent Office. *A List of Patents Granted by the United States from April 10, 1790, to December 31, 1836: With an Appendix Containing Reports on the Condition of the Patent-Office in 1823, 1830, and 1831.* Washington, DC, 1872.

Urban, Sylvanus. "History of the Late Comet." *Gentleman's Magazine* (London), November 1, 1759, 521–24.

"Valuable Secrets." *The Columbian Almanac, or The North American Almanac*, 1804. Wilmington, DE, 1803.

Van Dyke, Rachel. *To Read My Heart: The Journal of Rachel Van Dyke, 1810–1811.* Edited by Lucia McMahon and Deborah Schriver. Philadelphia: University of Pennsylvania Press, 2000.

Visitors Guide to the Centennial Exhibition and Philadelphia. Philadelphia: J. B. Lippincott, 1876.

Ward, H. A. *Catalog of Casts and Fossils from the Principal Museums of Europe and America with Short Descriptions and Illustrations.* Rochester, NY: Benton and Andrews, 1866.

Warne, Joseph A. *Phrenology in the Family: or, The Utility of Phrenology in Early Domestic Education.* Philadelphia: George W. Donohue, 1839.

Watson, Elkanah. *History of the Rise, Progress, and Existing Conditions of the Western Canals in the State of New York [. . .] Together with the Rise, Progress, and Existing State of Modern Agricultural Societies, on the Berkshire System, from 1807, to the Establishment of the Board of Agriculture in the State of New-York, January 10, 1820.* Albany, NY: D. Steele, 1820.

Weatherwise's Town and Country Almanac for 1785. Boston: Weeden and Barrett, 1784.

Withington, William. *Crystal Palace and the World's Fair in New York City*. Lawrence, MA: n.p., 1853.

Wright, Susanna. "Directions for the Management of Silk Worms. By the Late Mrs. S. Wright, of Lancaster-County, in Pennsylvania." *Philadelphia Medical and Physical Journal* 1 (1804): 103–7.

Secondary Sources

Ackerman, Silke. "Maths and Memory: Calendar Medals in the British Museum, Part 1." *Medal* 45 (2004): 3–44.

Allison, Robert J. *Stephen Decatur: American Naval Hero, 1779–1820*. Amherst: University of Massachusetts Press, 2005.

Andrews, J. Cutler. *The North Reports the Civil War*. Pittsburgh: University of Pittsburgh Press, 1955.

Anishanslin, Zara. *Portrait of a Woman in Silk: Hidden Histories of the British Atlantic World*. New Haven, CT: Yale University Press, 2016.

Appleby, Joyce Oldham. *Inheriting the Revolution: The First Generation of Americans*. Cambridge, MA: Harvard University Press, 2000.

Aspiz, Harold. "Phrenologizing the Whale." *Nineteenth-Century Fiction* 23, no. 1 (June 1968): 18–27.

Auerbach, Jeffrey A. *The Great Exhibition of 1851: A Nation on Display*. New Haven, CT: Yale University Press, 1999.

Autin, John C. *Chelsea Porcelain at Williamsburg*. Williamsburg, VA: Colonial Williamsburg Foundation, 1977.

Baatz, Simon. "'Squinting at Silliman': Scientific Periodicals in the Early American Republic, 1810–1833." *Isis* 82, no. 2 (June 1991): 223–44.

Baatz, Simon. *"Venerate the Plough": A History of the Philadelphia Society for Promoting Agriculture, 1785–1985*. Philadelphia: Philadelphia Society for Promoting Agriculture, 1985.

Bailyn, Bernard. "1776: A Year of Challenge—a World Transformed." *Journal of Law and Economics* 19, no. 3 (1976): 437–66.

Bathe, Greville, and Dorothy Bathe. *Oliver Evans: A Chronicle of Early American Engineering*. Philadelphia: Historical Society of Pennsylvania, 1935. Reprint, New York: Arno, 1972.

Bedini, Silvio A. *Benjamin Banneker*. Rancho Cordova, CA: Landmark Enterprises, 1972.

Bedini, Silvio A. *Thomas Jefferson, Statesman of Science*. New York: Macmillan, 1990.

Betts, Charles. "The Yale College Expedition of 1870." *Harper's Monthly Magazine*, October 1871, 663–71.

Blaisdell, Thomas C., and Peter Selz. *The American Presidency in Political Cartoons, 1776–1976*. Berkeley, CA: University Art Museum, 1976.

Blake, Nelson M. *Water for the Cities: A History of the Urban Water Supply Problem in the United States*. Syracuse, NY: Syracuse University Press, 1956.

Blecki, Catherine La Courreye, and Karin A. Wulf, eds. *Milcah Martha Moore's Book: A Commonplace Book from Revolutionary America*. University Park: Pennsylvania State University Press, 1997.

Bleiler, E. F. "From the Newark Steam Man to Tom Swift." *Extrapolation* 30, no. 2 (Summer 1989): 101–16.

Block, Eugene B. *Above the Civil War: The Story of Thaddeus Lowe, Balloonist, Inventor, Railway Builder.* Berkeley, CA: Howell-North, 1966.

Bode, Carl. *The American Lyceum.* New York: Oxford University Press, 1956.

Borneman, Walter R. *1812: The War That Forged a Nation.* New York: HarperCollins, 2004.

Bowman, Hank Wieand. *Pioneer Railroads.* Greenwich, CT: Fawcett, 1954.

Bramwell, Valerie, and Robert M. Peck. *All in the Bones: A Biography of Benjamin Waterhouse Hawkins.* Philadelphia: Academy of Natural Sciences of Philadelphia, 2008.

Branson, Susan. "Flora and Femininity: Gender and Botany in Early America." *Commonplace* 12, no. 2 (January 2012). http://commonplace.online/article/flora-femininity/.

Breen, T. H. "An Empire of Goods: The Anglicization of Colonial America, 1690–1776." *Journal of British Studies* 25, no. 4 (October 1986): 467–99.

Brigham, David R. *Public Culture in the Early Republic: Peale's Museum and Its Audience.* Washington, DC: Smithsonian Institution Press, 1995.

Brooks, Van Wyck. *The Flowering of New England.* New York: Dutton, 1936.

Brown, Chandos. "A Natural History of the Gloucester Sea Serpent: Knowledge, Power and Science in Antebellum America." *American Quarterly* 42, no. 3 (September 1990): 402–36.

Browne, C. A. "Elder John Leland and the Mammoth Cheshire Cheese." *Agricultural History* 18, no. 4 (October 1944): 145–53.

Burnard, Trevor. *Mastery, Tyranny, and Desire: Thomas Thistlewood and His Slaves in the Anglo-Jamaican World.* Chapel Hill: University of North Carolina Press, 2004.

Burrows, Edwin G. *The Finest Building in America: The New York Crystal Palace, 1853–1858.* New York: Oxford University Press, 2018.

Butterfield, Lyman H. "Elder John Leland, Jeffersonian Itinerant." *Proceedings of the American Antiquarian Society* 62, pt. 2 (1952): 155–242.

Carlson, C. Lennart. "Samuel Keimer, A Study in the Transit of English Culture to Colonial Pennsylvania." *Pennsylvania Magazine of History and Biography* 61, no. 4 (1937): 357–86.

Carrott, Richard G. "The Neo-Egyptian Style in American Architecture." *Antiques* 90, no. 4 (October 1966): 482–88.

Carter, Edward C., II, ed. *The Virginia Journals of Benjamin Henry Latrobe, 1795–1798.* Vol. 2, *1797–1798.* New Haven, CT: Yale University Press, 1977.

A Catalogue of Instruments and Models in the Possession of the American Philosophical Society. Compiled by Robert P. Multhauf, assisted by David Davies. Philadelphia: American Philosophical Society, 1961.

Cerami, Charles. *Benjamin Banneker: Surveyor, Astronomer, Publisher, Patriot.* New York: John Wiley & Sons, 2002.

Clark, Charles E. "The Newspapers of Provincial America." In *Three Hundred Years of the American Newspaper: Essays,* edited by John B. Hench, 367–89. Worcester, MA: American Antiquarian Society, 1991.

Cohen, Joanna. *Luxurious Citizens: The Politics of Consumption in Nineteenth-Century America.* Philadelphia: University of Pennsylvania Press, 2017.

Cohen, Norm. *Long Steel Railroad: The Railroad in American Folksong.* Urbana: University of Illinois Press, 2000.

Coleman, Earle Edson. "The Exhibition in the Palace: A Bibliographic Essay." *Bulletin of the New York Public Library* 64, no. 9 (September 1960): 459–78.

Comment, Bernard. *The Painted Panorama.* New York: Harry N. Abrams, 1999.

Connaughton, Michael E. "'Ballomania': The American Philosophical Society and Eighteenth-Century Science." *Journal of American Culture* 7 (Spring/Summer 1984): 71–74.

Cooke, Jacob E. "Tench Coxe, Alexander Hamilton, and the Encouragement of American Manufactures." *William and Mary Quarterly* 32, no. 3 (July 1975): 369–92.

Cooke, Jacob E. *Tench Coxe and the Early Republic.* Chapel Hill: University of North Carolina Press, 1978.

Coon, David L. "Eliza Lucas Pinckney and the Reintroduction of Indigo Culture in South Carolina." *Journal of Southern History* 42, no. 1 (February 1976): 61–76.

Copeland, David A. *Colonial American Newspapers: Character and Content.* Newark: University of Delaware Press, 1997.

Cordato, Mary Frances. "Toward a New Century: Women and the Philadelphia Centennial Exhibition, 1876." *Pennsylvania Magazine of History and Biography* 107, no. 1 (January 1983): 113–35.

Crouch, Tom D. *The Eagle Aloft: Two Centuries of the Balloon in America.* Washington, DC: Smithsonian Institution Press, 1983.

Dain, Bruce. *A Hideous Monster of the Mind: American Race Theory in the Early Republic.* Cambridge, MA: Harvard University Press, 2002.

Dalzell, Robert F. *American Participation at the Great Exhibition of 1851.* Amherst, MA: Amherst College Press, 1960.

Davies, J. D. *Phrenology: Fad and Science: A 19th-Century American Crusade.* New Haven, CT: Yale University Press, 1955.

Davis, Robyn Lily. *See* McMillin, Robyn Davis.

Davison, Nancy Reynolds. "E. W. Clay: American Political Caricaturist of the Jacksonian Era." PhD diss., University of Michigan, 1980.

Delbourgo, James A. *A Most Amazing Scene of Wonders: Electricity and Enlightenment in Early America.* Cambridge, MA: Harvard University Press, 2007.

Dilts, James D. *The Great Road: The Building of the Baltimore and Ohio, the Nation's First Railroad, 1828–1853.* Stanford, CA: Stanford University Press, 1993.

Dinsmoor, William. "Early American Studies of Mediterranean Archaeology." *American Philosophical Society Proceedings* 87, no. 1 (1943): 70–104.

Dircks, Henry. *Perpetuum Mobile; or, A History of the Search for Self-Motive Power, from the 13th to the 19th Century.* London: E. & F. N. Spon., 1861. Reprint, Amsterdam: B. M. Israël, 1968.

Dreisbach, Daniel L. "Mr. Jefferson, a Mammoth Cheese, and the 'Wall of Separation between Church and State': A Bicentennial Commemoration." *Journal of Church and State* 43, no. 4 (Autumn 2001): 725–45.

Dugatkin, Lee Alan. *Mr. Jefferson and the Giant Moose: Natural History in Early America.* Chicago: University of Chicago Press, 2009.

Dupree, Hunter. *Science in the Federal Government: A History of Policies and Activities.* Cambridge, MA: Harvard University Press, 1957.

Ekirch, Arthur Alphonse, Jr. *The Idea of Progress in America, 1815–1860*. New York: AMS, 1969.

Evans, Charles M. *The War of the Aeronauts: A History of Ballooning during the Civil War*. Mechanicsburg, PA: Stackpole Books, 2002.

Fabian, Ann. *The Skull Collectors: Race, Science, and America's Unburied Dead*. Chicago: University of Chicago Press, 2010.

Finger, Simon. *The Contagious City: The Politics of Public Health in Early Philadelphia*. Ithaca, NY: Cornell University Press, 2012.

Foner, Philip S. "Black Participation in the Centennial of 1876." *Phylon* 39, no. 4 (1978): 283–96.

Fredrickson, George M. *The Black Image in the White Mind: The Debate on Afro-American Character and Destiny, 1817–1914*. Middletown, CT: Wesleyan University Press, 1971.

Frick, Margaret. "From Palace to Parlor: Exhibition Display, Consumerism, and Cultivation of Taste at the New York Crystal Palace." In *New York Crystal Palace 1853*. Digital publication based on a 2017 Focus Gallery exhibition at Bard Graduate Center. Accessed May 1, 2018. http://crystalpalace.visualizingnyc.org/digital-publication.

Giberti, Bruno. *Designing the Centennial*. Lexington: University of Kentucky Press, 2002.

Gibson, Jane Mork, and Robert Wolterstorff. "The Fairmount Waterworks." *Philadelphia Museum of Art Bulletin* 84, nos. 360/361 (Summer 1988): 1–46.

Gish, Dustin, and Daniel Klinghard. *Thomas Jefferson and the Science of Republican Government: A Political Biography of Notes on the State of Virginia*. Cambridge: Cambridge University Press, 2017.

Gold, Susanna W. *The Unfinished Exhibition: Visualizing Myth, Memory, and the Shadow of the Civil War in Centennial America*. London: Routledge, 2017.

Goodman, Matthew. *The Sun and the Moon: The Remarkable True Account of Hoaxers, Showmen, Dueling Journalists, and Lunar Man-Bats in Nineteenth-Century New York*. New York: Basic Books, 2008.

Gorin, Roberta. "In Remembrance of Things Past: Souvenirs of the Crystal Palace." In *New York Crystal Palace 1853*. Digital publication based on a Focus Gallery exhibition at Bard Graduate Center, March 24–July 30, 2017. http://crystalpalace.visualizingnyc.org/digital-publication/in-remembrance-of-things-past-souvenirs-of-the-crystal-palace/.

Gottesman, Rita Susswein. *Arts and Crafts in New York, 1800–1804: Advertisements and News Items from New York City Newspapers*. New York: New-York Historical Society, 1965.

Goubert, Jean-Pierre. *The Conquest of Water: The Advent of Health in the Industrial Age*. Princeton, NJ: Princeton University Press, 1989.

Grant, A. Cameron. "George Combe and American Slavery." *Journal of Negro History* 45, no. 4 (October 1960): 259–69.

Greene, John C. "Science, Learning, and Utility: Patterns of Organization in the Early American Republic." In Oleson and Brown, *Pursuit of Knowledge*, 1–20.

Gribben, Alan. "Mark Twain, Phrenology and the 'Temperaments': A Study of Pseudoscientific Influence." *American Quarterly* 24, no. 1 (March 1972): 45–68.

Gronim, Sara Stidstone. *Everyday Nature: Knowledge of the Natural World in Colonial New York*. New Brunswick, NJ: Rutgers University Press, 2007.

Gronim, Sara Stidstone. "What Jane Knew: A Woman Botanist in the Eighteenth Century." *Journal of Women's History* 19, no. 3 (Fall 2007): 33–59.

Hardin, Laura Vookles. "Celebrating the Aqueduct: Pastoral and Urban Ideas." In *The Old Croton Aqueduct: Rural Resources Meet Urban Needs*, edited by Jeffrey Kroessler, unpaginated. Yonkers, NY: Hudson River Museum of Westchester, 1992.

Hedeen, Stanley. *Big Bone Lick: The Cradle of American Paleontology*. Lexington: University of Kentucky Press, 2008.

Hillway, Tyrus. "Melville's Use of Two Pseudo-Sciences." *Modern Language Notes* 64, no. 3 (March 1949): 145–50.

Hindle, Brooke. *The Pursuit of Science in Revolutionary America, 1735–1789*. Chapel Hill: University of North Carolina Press, 1956.

Hindle, Brooke. "The Underside of the Learned Society in New York, 1754–1854." In Oleson and Brown, *Pursuit of Knowledge*, 84–116.

Hindle, Brooke, and Steven Lubar. *Engines of Change: The American Industrial Revolution, 1790–1860*. Washington, DC: Smithsonian Institution Press, 1986.

Hopkins, Lisa. "Jane C. Loudon's *The Mummy!*: Mary Shelley Meets George Orwell and They Go in a Balloon to Egypt." *Cardiff Corvey: Reading the Romantic Text* 10 (June 2003). Accessed June 19, 2019. http://www.cf.ac.uk/encap/corvey/articles/cc10_n01.pdf.

Hyde, Ralph. *Panoramania!* London: Trefoil, 1988.

Jaffe, David. "The Ebenezers Devotion: Pre- and Post-Revolutionary Consumption in Rural New England." *New England Quarterly* 76, no. 2 (June 2003): 239–64.

James, Mary Ann. "Engineering an Environment for Change: Bigelow, Pierce and Early Practical Education at Harvard." In *Science at Harvard University: Historical Perspectives*, edited by Clark A. Elliott and Margaret W. Rossiter, 55–75. Bethlehem, PA: Lehigh University Press, 1992.

James, Reese D. *Old Drury of Philadelphia: A History of the Philadelphia Stage*. New York: Greenwood, 1968.

Jayne, Thomas Gordon. "The New York Crystal Palace: An International Exhibition of Goods and Ideas." Master's thesis, University of Delaware, 1990.

Jeppson, Patrice L. "Comets and Calendars." Talk presented as part of Explore Philadelphia's Hidden Past: A Pennsylvania Archaeology Month Celebration. Sponsored by the Philadelphia Archaeological Forum and Independence National Historical Park. Philadelphia, 2007.

Johns, Richard R. *Spreading the News: The American Postal System from Franklin to Morse*. Cambridge, MA: Harvard University Press, 1998.

Jones, Douglas A., Jr. *The Captive Stage: Performance and the Proslavery Imagination of the Antebellum North*. Ann Arbor: University of Michigan Press, 2014.

Jones, Howard. *Mutiny on the* Amistad: *The Saga of a Slave Revolt and Its Impact on American Abolition, Law, and Diplomacy*. New York: Oxford University Press, 1987.

Kaser, David. *A Book for a Sixpence: The Circulating Library in America*. Pittsburgh: Beta Phi Mu, 1980.

Kaufman, Matthew H. *Edinburgh Phrenological Society: A History*. Edinburgh: William Ramsay Henderson Trust, 2005.

Kierner, Cynthia A. *Inventing Disaster: The Culture of Calamity from the Jamestown Colony to the Johnstown Flood*. Chapel Hill: University of North Carolina Press, 2019.

Kieve, Jeffrey. *The Electric Telegraph: A Social and Economic History*. Newton Abbot, UK: David and Charles, 1973.

Klunder, Willard Carl. *Lewis Cass and the Politics of Moderation*. Kent, OH: Kent State University Press, 1996.

Koeppel, Gerard T. *Water for Gotham: A History*. Princeton, NJ: Princeton University Press, 2001.

Lepler, Jessica M. *The Many Panics of 1837: People, Politics, and the Creation of a Transatlantic Financial Crisis*. Cambridge: Cambridge University Press, 2013.

Linker, Jessica C. "The Fruits of Their Labor: Women's Scientific Practice in Early America, 1750–1860." PhD diss., University of Connecticut, 2017.

Lipman, Jean. *Rufus Porter Rediscovered: Artist, Inventor, Journalist, 1792–1884*. New York: C. N. Potter, 1980.

Looby, Christopher. "The Constitution of Nature: Taxonomy as Politics in Jefferson, Peale, and Bartram." *Early American Literature* 22, no. 3 (1987): 252–73.

Lynn, Catherine. *Wallpaper in America: From the Seventeenth Century to World War I*. New York: W. W. Norton, 1980.

Marceau, Henri. *William Rush: The First Native American Sculptor*. Philadelphia: Pennsylvania Museum of Art, 1937.

Marsh, Ben. "Silk Hopes in Colonial South Carolina." *Journal of Southern History* 78, no. 4 (November 2012): 807–54.

Marsh, Ben. *Unravelled Dreams: Silk and the Atlantic World, 1500–1840*. Cambridge: Cambridge University Press, 2020.

Martin, A. L. *Villain of Steam: A Life of Dionysius Lardner*. Carlow, Ireland: Tyndall Scientific, 2012.

McCarthy, Molly A. *The Accidental Diarist: A History of the Daily Planner in America*. Chicago: University of Chicago Press, 2013.

McCarthy, Molly A. "Redeeming the Almanac: Learning to Appreciate the iPhone of Early America." *Commonplace* 11, no. 1 (October 2010). http://commonplace.online/article/redeeming-the-almanac/.

McClellan, James E. *Science Reorganized: Scientific Societies in the Eighteenth Century*. New York: Columbia University Press, 1985.

McCoy, Janet Rice. "Dr. R. C. Rutherford, Phrenologist and Lecturer: His Public Humiliation by Matrimony." *Northwest Ohio Quarterly* 74 (Summer/Fall 2002): 152–66.

McKendrick, Neil. *The Birth of a Consumer Society: The Commercialization of Eighteenth-Century England*. Bloomington: Indiana University Press, 1982.

McMillin, Robyn Davis. "Science in the American Style, 1700–1800." PhD diss., University of Oklahoma, 2009.

Melish, Joanne Pope. *Disowning Slavery: Gradual Emancipation and "Race" in New England, 1780–1860*. Ithaca, NY: Cornell University Press, 1998.

Milroy, Elizabeth. *The Grid and the River: Philadelphia's Green Places, 1682–1876*. University Park: Penn State University Press, 2016.

Morgan, Jennifer L. *Laboring Women: Reproduction and Gender in New World Slavery.* Philadelphia: University of Pennsylvania Press, 2004.

Morton, Henry B. "The Redheffer Perpetual Motion Machine." *Journal of the Franklin Institute* (Philadelphia) 139 (1894–95): 246–51.

Mott, Frank Luther. *A History of American Magazines.* Vol. 1, *1741–1850.* Cambridge, MA: Belknap Press of Harvard University Press, 1930.

Murrin, John. "A Roof without Walls." In *Beyond Confederation: Origins of the Constitution and American National Identity*, edited by Richard R. Beeman, Stephen Botein, and Edward C. Carter II, 333–48. Williamsburg, VA: Omohundro Institute of Early American History and Culture, 1987.

Nadis, Fred. *Wonder Shows: Performing Science, Magic, and Religion in America.* New Brunswick, NJ: Rutgers University Press, 2005.

Naeve, Milo M. *John Lewis Krimmel: An Artist in Federal America.* Newark: University of Delaware Press, 1988.

Nester, William. *The Age of Jackson and the Art of American Power, 1815–1848.* Washington, DC: Potomac Books, 2013.

Nocks, Lisa. *The Robot: The Life Story of a Technology.* Westport, CT: Greenwood, 2006.

Noordung, Hermann. *The Problem of Space Travel: The Rocket Motor.* Washington, DC: National Aeronautics and Space Administration, 1995.

O'Connor, Ralph. *The Earth on Show: Fossils and the Poetics of Popular Science, 1802–1856.* Chicago: University of Chicago Press, 2007.

O'Connor, Thomas H. *The Athens of America: Boston, 1825–1845.* Amherst, MA: University of Massachusetts Press, 2006.

Oleson, Alexandra, and Sanborn C. Brown, eds. *The Pursuit of Knowledge in the Early American Republic: American Scientific and Learned Societies from Colonial Times to the Civil War.* Baltimore: Johns Hopkins University Press, 1976.

Ord-Hume, Arthur. *Perpetual Motion: The History of an Obsession.* New York: St. Martin's, 1977.

Palmquist, Peter E., and Thomas R. Kailbourn. *Pioneer Photographers from the Mississippi to the Continental Divide: A Biographical Dictionary, 1839–1865.* Stanford, CA: Stanford University Press, 2005.

Park, Benjamin E. *American Nationalisms: Imagining Union in the Age of Revolutions, 1783–1833.* New York: Cambridge University Press, 2018.

Parrish, Susan S. *American Curiosity: Cultures of Natural History in the Colonial British Atlantic World.* Chapel Hill: University of North Carolina Press, 2006.

Pasley, Jeffrey. "The Cheese and the Words: Popular Political Culture and Participatory Democracy in the Early American Republic." In *Beyond the Founders: New Approaches to the Political History of the Early American Republic*, edited by Jeffrey L. Pasley, Andrew W. Robertson, and David Waldstreicher, 31–54. Chapel Hill: University of North Carolina Press, 2004.

Pawley, Emily. *The Nature of the Future: Agriculture, Science, and Capitalism in the Antebellum North.* Chicago: University of Chicago Press, 2020.

Piola, Erika, and Jennifer Ambrose. "The First Fifty Years of Commercial Lithography in Philadelphia." In *Philadelphia on Stone: Commercial Lithography in Philadelphia, 1828–1878*, edited by Erika Piola, 1–47. University Park: Penn State University Press, 2012.

Pollack, John H., ed. *"The Good Education of Youth": Worlds of Learning in the Age of Franklin*. New Castle, DE: University of Pennsylvania Libraries and Oak Knoll Press, 2009.

Post, Robert C., ed. *1876: A Centennial Exhibition*. Washington, DC: Smithsonian Institution, 1976.

Powell, Richard J. "Cinqué: Antislavery Portraiture and Patronage in Jacksonian America." *American Art* 11, no. 3 (Autumn 1997): 48–73.

Power-Greene, Ousmane K. *Against Wind and Tide: The African American Struggle against the Colonization Movement*. New York: NYU Press, 2014.

Price, George R., and James Brewer Stewart. *To Heal the Scourge of Prejudice: The Life and Writings of Hosea Easton*. Amherst: University of Massachusetts Press, 1999.

Pyle, Gerald F. "The Diffusion of Cholera in the United Sates in the Nineteenth Century." *Geographical Analysis* 1, no. 1 (1969): 59–75.

Raven, James. *London Booksellers and American Customers: Transatlantic Literary Community and the Charleston Library Society, 1748–1811*. Columbia: University of South Carolina Press, 2002.

Raven, James. "Social Libraries and Library Societies in Eighteenth-Century North America." In *Institutions of Reading: The Social Life of Libraries in the United States*, edited by Thomas Augst and Kenneth Carpenter, 24–52. Amherst: University of Massachusetts Press, 2007.

Rawson, Michael. "The Nature of Water: Reform and the Antebellum Crusade for Municipal Water in Boston." *Environmental History* 9, no. 3 (July 2004): 411–35.

Ray, Angela G. *The Lyceum and Public Culture in the Nineteenth-Century United States*. East Lansing: Michigan State University Press, 2005.

Rediker, Marcus. *The Amistad Rebellion: An Atlantic Odyssey of Slavery and Freedom*. New York: Viking, 2012.

Reilly, Elizabeth Carroll, and David D. Hall. "Customers and the Market for Books." In *The Colonial Book in the Atlantic World*, edited by David Hall and Hugh Amory, 387–98. Chapel Hill: University of North Carolina Press in association with the American Antiquarian Society, 2007.

Reinhold, Meyer. "The Quest for 'Useful Knowledge' in Eighteenth-Century America." *Proceedings of the American Philosophical Society* 119, no. 2 (April 16, 1975): 108–32.

Richardson, Edgar P., Brooke Hindle, and Lillian B. Miller. *Charles Willson Peale and His World*. New York: Abrams, 1983.

Riegel, Robert E. "The Introduction of Phrenology to the United States." *American Historical Review* 39, no. 1 (October 1933): 73–78.

Rigal, Laura. "Peale's Mammoth." In *American Iconology: New Approaches to Nineteenth-Century Art and Literature*, edited by David C. Miller, 18–38. New Haven, CT: Yale University Press, 1993.

Risley, Ford. *Civil War Journalism*. Santa Barbara, CA: Praeger, 2012.

Roberts, Christopher M. "The Water Works: A Place 'Wondrous to Behold.'" *Delaware River Basin Commission Annual Report*. West Trenton, NJ: DRBC, 1992. Accessed June 26, 2016. https://www.nj.gov/drbc/about/public/annual-reports.html.

Roediger, David R. *The Wages of Whiteness: Race and the Making of the American Working Class*. London: Verso, 1991.

Rolt, L. T. C. *The Aeronauts: A History of Ballooning, 1783–1903*. New York: Walker, 1966.

Rosen, William. *The Most Powerful Idea in the World: A Story of Steam, Industry and Invention*. Chicago: University of Chicago Press, 2010.

Rosenberg, Charles E. *The Cholera Years*. Chicago: University of Chicago Press, 1962.

Rossiter, Margaret W. "Benjamin Silliman and the Lowell Institute: The Popularization of Science in Nineteenth-Century America." *New England Quarterly* 44 (1971): 602–26.

Ruffin, J. Rixby. "'Urania's Dusky Vails': Heliocentrism in Colonial Almanacs, 1700–1735." *New England Quarterly* 70, no. 2 (June 1997): 306–13.

Rydell, Robert W. *All the World's a Fair: Visions of Empire at American International Expositions, 1876–1916*. Chicago: University of Chicago Press, 1984.

Sappol, Michael. *A Traffic of Dead Bodies: Anatomy and Embodied Social Identity in Nineteenth-Century America*. Princeton, NJ: Princeton University Press, 2002.

Schaffer, Simon. "Natural Philosophy and Public Spectacle in the Eighteenth Century." *History of Science* 21, no. 1 (1983): 1–43.

Schaffer, Simon. "The Show That Never Ends: Perpetual Motion in the Early Eighteenth Century." *British Journal for the History of Science* 28, pt. 2, no. 97 (June 1995): 157–89.

Schiebinger, Londa L., and Claudia Swan, eds. *Colonial Botany: Science, Commerce, and Politics in the Early Modern World*. Philadelphia: University of Pennsylvania Press, 2005.

Scott, Donald M. "The Popular Lecture and the Creation of a Public in Mid-Nineteenth-Century America." *Journal of American History* 66, no. 4 (March 1980): 791–809.

Sellers, Charles Coleman. *Mr. Peale's Museum: Charles Willson Peale and the First Popular Museum of National Science and Art*. New York: W. W. Norton, 1980.

Semonin, Paul. *American Monster: How the Nation's First Prehistoric Creature Became a Symbol of National Identity*. New York: NYU Press, 2000.

Shapiro, Henry D. "The Western Academy of Natural Sciences of Cincinnati and the Structure of Science in the Ohio Valley, 1810–1850." In Oleson and Brown, *Pursuit of Knowledge*, 219–47.

Sharnoff, Philip, ed. *Annals of Cleveland, 1818–1935*. Vol. 41. Cleveland: Cleveland Public Library, 1937.

Sinclair, Bruce. *Philadelphia's Philosopher Mechanics: A History of the Franklin Institute, 1824–1865*. Baltimore: Johns Hopkins University Press, 1974.

Sinclair, Bruce. "Science, Technology, and the Franklin Institute." In Oleson and Brown, *Pursuit of Knowledge*, 194–207.

Skiba, Bob. "Dance and the Urban Landscape in the 19th Century." *Philadelphia Dance History Journal*, January 25, 2012. https://philadancehistoryjournal.wordpress.com/tag/fairmount-water-works/.

Smith, Billy G. *Ship of Death: A Voyage That Changed the Atlantic World*. New Haven, CT: Yale University Press, 2013.

Smith, Carl S. *City Water, City Life: Water and the Infrastructure of Ideas in Urbanizing Philadelphia, Boston, and Chicago*. Chicago: University of Chicago Press, 2013.

Smith, Kristen M., and Jennifer L. Gross. *The Lines Are Drawn: Political Cartoons of the Civil War*. Athens, GA: Hill Street, 1999.

Smith, Murphy D. *A Museum: The History of the Cabinet of Curiosities of the American Philosophical Society*. Philadelphia: American Philosophical Society, 1996.

Soll, David. *Empire of Water: An Environmental and Political History of the New York City Water Supply*. Ithaca, NY: Cornell University Press, 2013.

Stack, David. *Queen Victoria's Skull: George Combe and the Mid-Victorian Mind*. London: Hambledon Continuum, 2008.

Stafford, Barbara Maria. *Artful Science: Enlightenment Entertainment and the Eclipse of Visual Education*. Cambridge, MA: MIT Press, 1994.

Standage, Tom. *The Victorian Internet*. London: Weidenfield & Nicolson, 1998.

Stearns, Raymond Phineas. *Science in the British Colonies of America*. Urbana: University of Illinois Press, 1970.

Steen, Ivan D. "America's First World's Fair: The Exhibition of the Industry of All Nations at New York's Crystal Palace, 1853–1854." *New York Historical Society Quarterly* 47 (July 1963): 257–87.

Stern, Madeleine B. *Heads and Headlines: The Phrenological Fowlers*. Norman: University of Oklahoma Press, 1971.

Stewart, James Brewer. "The Emergence of Racial Modernity and the Rise of the White North, 1790–1840." *Journal of the Early Republic* 18, no. 2 (Summer 1998): 181–217.

Stoehr, Taylor. *Hawthorne's Mad Scientists: Pseudoscience and Social Science in Nineteenth-Century Life and Letters*. Hamden, CT: Archon Books, 1978.

Strang, Cameron. "Scientific Instructions and Native American Linguistics in the Imperial United States: The Department of War's 1826 Vocabulary." *Journal of the Early Republic* 37 (Fall 2017): 399–427.

Sutcliffe, Andrea. *Steam: The Untold Story of America's First Great Invention*. New York: Palgrave Macmillan, 2004.

Sutton, Geoffrey. *Science for a Polite Society: Gender, Culture, and the Demonstration of Enlightenment*. New York: HarperCollins, 1995.

Thompson, D. Dodge. "The Public Work of William Rush: A Case Study in the Origins of American Sculpture." In *William Rush, American Sculptor*, edited by Richard D. Boyle, 31–46. Philadelphia: Pennsylvania Academy of Fine Arts, 1982.

Thomson, Keith. *The Legacy of the Mastodon: The Golden Age of Fossils in America*. New Haven, CT: Yale University Press, 2008.

Tomlinson, Stephen. *Head Masters: Phrenology, Secular Education, and Nineteenth-Century Social Thought*. Tuscaloosa: University of Alabama Press, 2005.

Tomlinson, Stephen. "Phrenology, Education and the Politics of Human Nature: The Thought and Influence of George Combe." *History of Education* 26, no. 1 (March 1997): 1–22.

Trask, Kerry A. *Black Hawk: The Battle for the Heart of America*. New York: Henry Holt, 2006.

Van Horn, Jennifer. *The Power of Objects in Eighteenth-Century British America*. Chapel Hill: published for the Omohundro Institute of Early American History and Culture, Williamsburg, Virginia, by the University of North Carolina Press, 2017.

Van Wyhe, John. "The Diffusion of Phrenology through Public Lecturing." In *Science in the Marketplace: Nineteenth-Century Sites and Experiences*, edited by Aileen Fyfe and Bernard Lightman, 60–96. Chicago: University of Chicago Press, 2007.

Van Wyhe, John. *Phrenology and the Origins of Victorian Scientific Naturalism*. Farnham, UK: Ashgate, 2004.

Walsh, Anthony A. "The American Tour of Dr. Spurzheim." *Journal of the History of Medicine and Allied Sciences* 27, no. 2 (1972): 187–205.

Walsh, Anthony A. "George Combe: A Portrait of a Heretofore Generally Unknown Behaviorist." *Journal of the History of the Behavioral Sciences* 7, no. 3 (July 1971): 269–78.

Walsh, Anthony A. "Phrenology and the Boston Medical Community in the 1830s." *Bulletin of the History of Medicine* 50 (1976): 261–73.

Warner, Deborah J. "Women Inventors at the Centennial." In *Dynamos and Virgins Revisited: Women and Technological Change in History*, edited by Martha Moore Trescott, 102–19. Metuchen, NJ: Scarecrow, 1979.

Warner, Deborah J. "The Women's Pavilion." In *Post, 1876*, 165–73.

Warren, Edward. *The Life of John Collins Warren, M.D.: Compiled Chiefly from His Autobiography and Journals*. Vol. 2. Boston: Ticknor and Fields, 1860.

Webster, Elizabeth E. "American Science and the Pursuit of 'Useful Knowledge' in the Polite Eighteenth Century, 1750–1806." PhD diss., University of Notre Dame, 2010.

Weigley, Russell F., ed. *Philadelphia: A 300-Year History*. 2nd ed. New York: W. W. Norton, 1982.

Westcott, Thompson. *The Life of John Fitch, the Inventor of the Steamboat*. Philadelphia: J. B. Lippincott, 1857.

Wilkinson, Ronald Sterne. "Poe's 'Balloon-Hoax' Once More." *American Literature* 32, no. 3 (November 1960): 313–17.

Winterer, Caroline. *Culture of Classicism: Ancient Greece and Rome in American Intellectual Life, 1780–1910*. Baltimore: Johns Hopkins University Press, 2001.

Woodhouse, Anne Felicity. "Nicholas Biddle in Europe, 1804–1807." *Pennsylvania Magazine of History and Biography* 103 (January 1979): 3–33.

Wulf, Andrea. *Chasing Venus: The Race to Measure the Heavens*. New York: Knopf, 2012.

Wulf, Karin A. Introduction to *Milcah Martha Moore's Book: A Commonplace Book from Revolutionary America*, 1–58. Edited by Catherine La Courreye Blecki and Karin A. Wulf. University Park: Pennsylvania State University Press, 1997.

Yamin, Rebecca. *Digging in the City of Brotherly Love*. New Haven, CT: Yale University Press, 2008.

Yokota, Kariann Akemi. *Unbecoming British: How Revolutionary America Became a Postcolonial Nation*. Oxford: Oxford University Press, 2011.

York, Neil Longley. *Mechanical Metamorphosis: Technological Change in Revolutionary America*. Westport, CT: Greenwood, 1985.

INDEX

Page numbers in italics refer to figures.